A word-cloud of authors' surnames arranged in the shape of a book/figure:

Defoe, Abbott, Marx, Hardy, Melville, Montaigne, Machiavelli, Haggard, Chesterton, Cooper, Emerson, Joyce, Austen, Hugo, Grimm, Eliot, Stoker, Christie, Carroll, Maupassant, Byron, Molière, Wilde, Garnett, Goethe, Cotton, Fitzgerald, Einstein, Hawthorne, Engels, Schiller, Smith, Kafka, Hall, Baum, Leslie, Dumas, Henry, Kipling, Doyle, Dostoyevsky, Willis, Flaubert, Nietzsche, Turgenev, Balzac, Stockton, Vatsyayana, Crane, Burroughs, Verne, Curtis, Tocqueville, Gogol, Vinci, Homer, Widger, Tolstoy, Whitman, Busch, Darwin, Thoreau, Potter, Freud, Zola, Lawrence, Twain, Scott, Plato, Harte, Kant, Jowett, Stevenson, Dickens, Hesse, Burton, Andersen, Cervantes, London, Descartes, Voltaire, Poe, Aristotle, Wells, Hale, James, Hastings, Cooke, Bunner, Shakespeare, Richter, Chambers, Irving, Doré, da, Benedict, Alcott, Dante, Shaw, Pushkin, Swift, Chekhov, Wodehouse, Newton

tredition

tredition was established in 2006 by Sandra Latusseck and Soenke Schulz. Based in Hamburg, Germany, tredition offers publishing solutions to authors and publishing houses, combined with worldwide distribution of printed and digital book content. tredition is uniquely positioned to enable authors and publishing houses to create books on their own terms and without conventional manufacturing risks.

For more information please visit: www.tredition.com

TREDITION CLASSICS

This book is part of the TREDITION CLASSICS series. The creators of this series are united by passion for literature and driven by the intention of making all public domain books available in printed format again - worldwide. Most TREDITION CLASSICS titles have been out of print and off the bookstore shelves for decades. At tredition we believe that a great book never goes out of style and that its value is eternal. Several mostly non-profit literature projects provide content to tredition. To support their good work, tredition donates a portion of the proceeds from each sold copy. As a reader of a TREDITION CLASSICS book, you support our mission to save many of the amazing works of world literature from oblivion. See all available books at www.tredition.com.

Project Gutenberg

The content for this book has been graciously provided by Project Gutenberg. Project Gutenberg is a non-profit organization founded by Michael Hart in 1971 at the University of Illinois. The mission of Project Gutenberg is simple: To encourage the creation and distribution of eBooks. Project Gutenberg is the first and largest collection of public domain eBooks.

Nelson's Home Comforts
Thirteenth Edition

Mary Hooper

Imprint

This book is part of TREDITION CLASSICS

Author: Mary Hooper
Cover design: Buchgut, Berlin – Germany

Publisher: tradition GmbH, Hamburg - Germany
ISBN: 978-3-8472-1549-3

www.tredition.com
www.tredition.de

Copyright:
The content of this book is sourced from the public domain.

The intention of the TREDITION CLASSICS series is to make world literature in the public domain available in printed format. Literary enthusiasts and organizations, such as Project Gutenberg, worldwide have scanned and digitally edited the original texts. tredition has subsequently formatted and redesigned the content into a modern reading layout. Therefore, we cannot guarantee the exact reproduction of the original format of a particular historic edition. Please also note that no modifications have been made to the spelling, therefore it may differ from the orthography used today.

NELSON'S

HOME COMFORTS.

THIRTEENTH EDITION.

REVISED AND ENLARGED
By MARY HOOPER,
AUTHOR OF "LITTLE DINNERS," "EVERY-DAY MEALS,"
"COOKERY FOR INVALIDS," *ETC. ETC.*

London:
G. NELSON, DALE & CO., LIMITED,
14, DOWGATE HILL.
1892.

ANY OF
NELSON'S SPECIALITIES
MENTIONED IN THIS BOOK
MAY BE OBTAINED FROM
W. CHAPLIN & SONS,
19 & 20, WATERLOO PLACE,
SOUTHAMPTON.

PLEASE SEND, S.W.R.

They are also Sold by Grocers, Chemists, Italian Warehousemen, etc., throughout the World. Should any difficulty be experienced in obtaining them, kindly send the name and address of your Grocer, and we will at once communicate with him.

TRADE 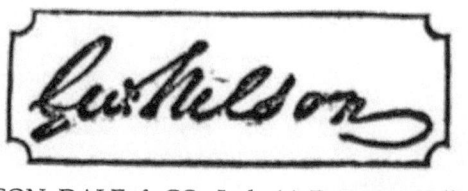 MARK.

G. NELSON, DALE, & CO., Ltd., 14, Dowgate Hill, London.

NELSON'S SPECIALITIES.

PATENT OPAQUE GELATINE.
In packets, from 6d. to 7s. 6d.
CITRIC ACID.
In 3d. packets. For use with the Gelatine.
ESSENCE OF LEMON, ALMONDS, & VANILLA.
In graduated bottles, 8d.
FAMILY JELLY BOXES.
7s. 6d. each.
Containing sufficient of the above materials for 12 quarts of Jelly.
BOTTLED WINE JELLIES (Concentrated).
CALF'S FOOT, LEMON, SHERRY, PORT, ORANGE, AND CHERRY.
Quarts, 2s. 6d.; Pints, 1s. 4d.; Half-pints, 9d.
TABLET JELLIES.
ORANGE, LEMON, CALF'S FOOT, CHERRY, RASPBERRY, VANILLA, PORT, SHERRY, ETC. Quarts, 9d.; Pints, 6d.; Half-pints, 3d.
WINE TABLET JELLIES.
PORT, SHERRY, ORANGE. Pints only, 9d.
PATENT REFINED ISINGLASS.
In 1s. packets.
GELATINE LOZENGES. LIQUORICE LOZENGES.
In Ornamental Tins, 6d.
JELLY-JUBES.
A most agreeable and nourishing Sweetmeat.
EXTRACT OF MEAT.
For Soups, Gravies, etc. In ounce packets, 4d.
PURE BEEF TEA.
In half-pint packets, 6d.

SOUPS.

Beef and Carrots	
Beef and Celery	In pint packets,
Beef and Onion	6d. each.
Mulligatawny	
Beef, Peas, and Vegetables	In quart packets,
Beef, Lentils, and Vegetables	6d. each.

Penny Packets of Soup for charitable purposes.

EGG ALBUMEN.
For clearing Jelly or Soup.
In boxes containing 12 packets, 9d. per box.

G. NELSON, DALE, & CO., Ltd., 14, Dowgate Hill, London.

LITTLE DINNERS,
How to serve them with Elegance and Economy.

By Mary Hooper.

Twenty-second Edition. Crown 8vo, cloth, price 2s. 6d.

"Shows us how to serve up a 'little dinner,' such as a philosopher might offer a monarch—good, varied, in good taste, and cheap. Exactly what the young English wife wishes to know, and what the ordinary cookery book does not teach her." — *Queen.*

EVERY-DAY MEALS,
Being Economic and Wholesome Recipes for Plain Dinners, Breakfasts, Luncheons, and Suppers.

By Mary Hooper.

Eighth Edition. Crown 8vo, cloth, price 2s. 6d.

"Our already deep obligations to Miss Hooper are weightily increased by this excellent and practical little book. The recipes for little dishes are excellent, and so clearly worded that presumptuous man instantly believes, on reading them, that he could descend into the kitchen and 'toss up' the little dishes without any difficulty." — *Spectator.*

COOKERY FOR INVALIDS,
For Persons of Delicate Digestion, and for Children.

By Mary Hooper.

Sixth Edition. Crown 8vo, cloth, price 2s. 6d.

"An epicure might be content with the little dishes provided by Miss Hooper; but, at the same time, the volume fills the utmost extent of promise held out in the title-page." — *Pall Mall Gazette.*

LONDON: KEGAN PAUL, TRENCH, & CO.

CONTENTS.

Preface
Bottled Jellies
Tablet Jellies
Lemon Sponge
Citric Acid and Pure Essence of Lemon
Pure Essence of Almonds and Vanilla
Gelatine Lozenges
Jelly-Jubes
Licorice Lozenges
Albumen
Extract of Meat
Soups
Beef Tea
New Zealand Mutton
Tinned Meats
Gelatine
Soups
Little Dishes of Fish
Little Dishes of Meat
Puddings
Jellies
Creams
Cakes
Beverages
Macaroni, etc.
Hints on Housekeeping
New Zealand Frozen Mutton

NELSON'S HOME COMFORTS.

PREFACE.

In presenting our friends and the public with the thirteenth edition of our "Home Comforts," we have the pleasure to remark that so greatly has the book been appreciated, that the large number of FIVE HUNDRED THOUSAND copies has been called for. The value of the Jubilee Edition was enhanced by some new recipes; these are repeated in the present edition, to which, also, some valuable additions have been made. Since the introduction of our Gelatine by the late Mr. G. Nelson, more than fifty years ago, we have considerably enlarged our list of specialities, and we have gratefully to acknowledge the public favour accorded to us.

Among those of our preparations which have met with so much appreciation and success, we would cite the following:

Nelson's Bottled Jellies.—It is sometimes so difficult, if not impossible, to have a first-class jelly made in private kitchens, that we venture to think our Bottled Jellies will be highly appreciated by all housekeepers. It is not too much to say that a ready-made jelly of the highest quality, and of the [8] best and purest materials, requiring only the addition of hot water, is now, for the first time, supplied. Careful experiments, extending over a long period of time, have been required to bring this excellent and very useful preparation to its present state of perfection, and it is confidently asserted that no home-made jelly can surpass it in purity, brilliancy, or delicacy of flavour. All that is necessary to prepare the jelly for the table is to dissolve it by placing the bottle in hot water, and then to add the given quantity of water to bring it to a proper consistency. It is allowed to stand until on the point of setting, and is then put into a mould.

Nelson's Calf's Foot, Lemon, Port, Sherry, Orange, and Cherry Jellies are now to be had of all first-class grocers, and are put up in bottles each containing sufficient of the concentrated preparation to make a quart, pint, or half-pint.

Nelson's Tablet Jellies are recommended for general use, are guaranteed of the purest and best materials, and are flavoured with the finest fruit essences. The Tablet Jellies are of so moderate a price as to be within the reach of all classes, and can be used as an everyday addition to the family bill of fare. They are not, however, intended as a substitute for high-class jellies, whether bottled or home-made.

The Tablet Jellies used as directed in the recipes make, in a few minutes, creams of a most delicate [9] kind, remarkable for smoothness of texture and fine flavour.

Nelson's Port, Sherry, and Orange Wine Tablet Jellies have now been added to the list.

Nelson's Lemon Sponge, supplied in tins, is a delicious novelty, and will be found to surpass any that can be made at home.

Nelson's Citric Acid and Pure Essence of Lemon.—In order to save the trouble of putting jelly through a strainer when required for invalids, we have introduced our Citric Acid and Essence of Lemon, and by their use a jelly clear enough for all ordinary purposes is made in a few minutes.

Lemonade and other beverages can be quickly made, and with less expense than by any other method, by using Nelson's Citric Acid and Essence of Lemon, and for these recipes are given. Delicious beverages are also made with Nelson's Bottled Jellies, see page 93.

Nelson's Pure Essence of Almonds and Vanilla.—These Extracts, like the Essence of Lemon, will be found of superior strength and flavour, and specially adapted for the recipes in this book.

Nelson's Gelatine Lozenges are not only a delicious sweetmeat, but most useful as voice lozenges, or in cases of sore or irritable throat. The flavour is very delicate and refreshing. Dissolved in water they make a useful beverage, and also a jelly suitable for children and invalids.

[10] **Nelson's Jelly-Jubes** will be found most agreeable and nourishing sweetmeats, deliciously flavoured with fruit essences. They can be used as cough lozenges, will be found soothing for delicate throats, are useful for travellers, and may be freely given to children.

Nelson's Licorice Lozenges are not only a favourite sweetmeat, but in cases of throat irritation and cough are found to be soothing and curative.

Nelson's Albumen is the white of eggs carefully dried and prepared, so that it will keep for an indefinite length of time. It is useful for any purpose to which the white of egg is applied, and answers well for clearing soup and jelly. When required for use, the albumen is soaked in cold water and whisked in the usual way.

Nelson's Extract of Meat.—The numerous testimonials which have been received as to the excellence of this preparation, as well as the great and universal demand for it, have afforded the highest satisfaction to us as the manufacturers, and have enabled us to offer it with increased confidence to the public. It is invaluable, whether for making soup or gravy, or for strengthening or giving flavour to many dishes; and it is not only superior to, but far cheaper than, any similar preparation now before the public.

Now that clear soup is so constantly required, and [11] a thing of every-day use, Nelson's Extract of Meat will be found a great boon. With the addition of a little vegetable flavouring, a packet of the Extract will make a pint of soup as good and as fine as that produced, at much labour and expense, from fresh meat. With a judicious use of the liquor derived from boiling fowls, rabbits, and fresh meat, an endless variety of soup may be made, by the addition of Nelson's Extract of Meat. Some recipes are given by which first-class soups can be prepared in a short time, at a very small cost, and with but little trouble. It may be as well to say that soaking for a few minutes in cold water facilitates the solution of the Extract of Meat.

Nelson's Soups are deserving of the attention of every housekeeper, for they combine all the elements of good nourishment, have an excellent flavour, both of meat and vegetables, are prepared by merely boiling the contents of a packet for fifteen minutes, and are so cheap as to be within everybody's means. Penny packets of these soups, for charitable purposes, will be found most useful and nourishing.

Those who have to cater for a family know how often a little soup will make up a dinner that would otherwise be insufficient; yet because of the time and trouble required in the preparation, it is impossible to have it. In a case like this, or when a supplementary dish is unexpectedly required, Nelson's Soups are most useful. Although these Soups are all that can be desired, made with water according to [12] the directions given with each packet, they can be utilised with great advantage for strengthening household stock.

For instance, the liquor in which a leg of mutton has been boiled, or of pork, if not too salt, can be at once, by using a packet or two of Nelson's Soup, converted into a delicious and nourishing soup, and at a cost surprisingly small. Or the bones of any joint can be made into stock, and, after all the fat has been skimmed off, have a packet of Nelson's Soup added, in the same manner as in the directions.

Nelson's Beef Tea will be found of the highest value, supplying a cup of unequalled nourishment, combining all the constituents of fresh beef. No other preparation now before the public contains that most important element, albumen, in a soluble form, as well as much of the fibrin of the meat. This Beef Tea is also generally relished by invalids, and merely requires to be dissolved in boiling water.

New Zealand Mutton.—For information respecting this meat, and the great advantage as well as economy of its use, see page 119.

Nelson's Tinned Meats, known as the "Tomoana Brand," are prepared at the works of Nelson Bros., Limited, Hawke's Bay, New Zealand, from the finest cattle of the country. Messrs. Nelson specially recommend their "Pressed Mutton and Green Peas," "Haricot Mutton," and "Pressed [13] Corned Mutton." The "Stewed Kidneys" will be found of a quality superior to any articles of the kind now in the market, while the price places them within the reach of all classes of consumers.

Nelson's Gelatine having now been favourably known all over the world for more than half a century, it is unnecessary to do more than observe that our efforts are constantly directed to supplying a perfectly pure article, always of the same strength and quality. When Russian isinglass was first introduced into this country, the prejudices against its use on the part of our great-grandmothers were violent and extreme; for those worthy ladies would not believe that some unfamiliar substance, of the origin of which they were either ignorant or doubtful, could form an efficient substitute for the well-known calves' feet and cow-heels, from which they had always been in the habit of making their jellies and blanc-manges. By degrees, however, the Gelatine made its way, and at length superseded the old system entirely; and its popularity is demonstrated by the fact that the works at Emscote, near Warwick, cover nearly five acres.

N.B.—It is necessary to call attention to the fact that in all the following recipes in which Nelson's Gelatine and Specialities are used, the quantities are calculated for *their manufactures only*, the quality and strength of which may be relied upon for uniformity. [14]

NELSON'S HOME COMFORTS.

SOUPS.

BEEF AND ONION SOUP.

A pint of very good soup can be made by following the directions which accompany each tin of Nelson's Beef and Onion Soup, viz. to soak the contents in a pint of cold water for fifteen minutes, then place over the fire, stir, and boil for fifteen minutes. It is delicious when combined with a tin of Nelson's Extract of Meat, thus producing a quart of nutritious and appetising soup.

NELSON'S MULLIGATAWNY SOUP.

Soaked in cold water for a quarter of an hour, and then boiled for fifteen minutes, Nelson's Mulligatawny Soup is very appetising and delicious. It should be eaten with boiled rice; and for those who like the soup even hotter than that in the above preparation, the accompanying rice may be curried. In either case the rice should be boiled so that each grain should be separate and distinct from the rest.

[15]

BEEF, LENTIL, AND VEGETABLE SOUP.

Pour one quart of boiling water upon the contents of a tin of Nelson's Soup of the above title, stirring briskly. The water must be boiling. A little seasoning of salt and pepper may be added for accustomed palates. This soup is perfectly delicious if prepared as follows: Cut two peeled onions into quarters, tie them in a muslin bag, and let the soup boil for twenty minutes with them. Take out the bag before serving the soup.

BEEF, PEA, AND VEGETABLE SOUP.

The directions printed on each packet of Nelson's Beef, Pea, and Vegetable Soup produce a satisfactory soup, but even this may be improved by the addition of the contents of a tin of Nelson's Extract of Meat and a handful of freshly-gathered peas. It is perhaps not generally known that pea-pods, usually thrown away as useless, impart a most delicious flavour to soup if boiled fast for two or three hours in a large saucepan, strained, and the liquor added to the soup, stock, or beef tea.

BEEF TEA AS A SOLID.

Soak the contents of a tin of Nelson's Beef Tea in a gill of water for ten minutes. Add to this the third of an ounce packet of Nelson's Gelatine, which has been soaked for two or three hours in half-a-pint of cold water. Put the mixture in a stewpan, and stir until it reaches boiling-point. Then put it into a [16] mould which has been rinsed with cold water. When thoroughly cold, this will turn out a most inviting and extremely nutritious dish.

CLEAR VERMICELLI SOUP.

Boil two minced onions in a quart of the liquor in which a leg of mutton has been boiled, skim well, and when the vegetables are tender strain them out. Pass the soup through a napkin, boil up, skim thoroughly, and when clear add the contents of a tin of Nelson's Extract of Meat, stirring until dissolved.

Boil two ounces of vermicelli paste in a pint of water until tender. Most shapes take about ten minutes. Take care that the water boils when you throw in the paste, and that it continues to do so during all the time of cooking, as that will keep the paste from sticking together. When done, drain it in a strainer, put it in the tureen, and pour the soup on to it.

SOUP JULIENNE.

Wash and scrape a large carrot, cut away all the yellow parts from the middle, and slice the red outside of it an inch in length, and the eighth of an inch thick. Take an equal quantity of turnip and three small onions, cut in a similar manner. Put them in a stewpan with two ounces of butter and a pinch of powdered sugar; stir over the fire until a nice brown colour, then add a quart of water and a teaspoonful of salt, and let all simmer together gently for two [17] hours. When done skim the fat off very carefully, and ten minutes before serving add the contents of a tin of Nelson's Extract of Meat, and a cabbage-lettuce cut in shreds and blanched for a minute in boiling water; simmer for five minutes and the soup will be ready. Many cooks, to save time and trouble, use the preserved vegetables, which are to be had in great perfection at all good Italian warehouses.

BROWN RABBIT SOUP CLEAR.

Fry a quarter of a pound of onions a light brown; mince a turnip and carrot and a little piece of celery; boil these until tender in three pints of the liquor in which a rabbit has been boiled, taking care to remove all scum as it rises; strain them out, and then pass the soup through a napkin. The soup should be clear, or nearly so, but if it is not, put it in a stewpan, boil and skim until bright; then throw in the contents of a tin of Nelson's Extract of Meat, soaked for a few minutes; stir until dissolved; add pepper and salt to taste.

HARE SOUP.

Half roast a hare, and, having cut away the meat in long slices from the backbone, put it aside to make an *entrée*. Fry four onions; take a carrot, turnip, celery, a small quantity of thyme and parsley, half-a-dozen peppercorns, a small blade of mace, some bacon-bones or a slice of lean ham, with the body of the hare cut up into small pieces; put all in two quarts of water with a little salt. When you have [18] skimmed the pot, cover close and allow it to boil gently for three hours, then strain it; take off every particle of fat, and having allowed the soup to boil up, add the contents of a tin of Nelson's Extract of Meat, and thicken it with a dessertspoonful of potato-flour; stir in two lumps of sugar, a glass of port wine, and season if necessary.

MULLIGATAWNY SOUP.

English cooks generally err in making both mulligatawny and curries too hot. It is impossible to give the exact quantity of the powder, because it varies so much in strength, and the cook must therefore be guided by the quality of her material. Mulligatawny may be made cheaply, and be delicious. The liquor in which meat or fowl has been boiled will make a superior soup, and fish-liquor will answer well. Slice and fry brown four onions, quarter, but do not peel, four sharp apples; boil them in three pints of stock until tender, then rub through a sieve to a pulp. Boil this up in the soup, skimming well; add the contents of a tin of Nelson's Extract of Meat, and stir in two ounces of flour and the curry-powder, mixed smooth in half-a-pint of milk. Any little pieces of meat, fowl, game, or fish may be added as an improvement to the soup. Just before serving taste that the soup is well-flavoured; add a little lemon-juice or vinegar.

THIN MULLIGATAWNY SOUP.

To a quart of the liquor in which a fresh haddock has been boiled, add half-a-pint of water in which [19] onions have been boiled. Stir into this, after it has been skimmed, and whilst boiling, the contents of a tin of Nelson's Extract of Meat, and a teaspoonful of curry-powder; let it boil up; add the juice of half a lemon and serve.

BROWN ARTICHOKE SOUP.

Wash, peel, and cut into slices about half-an-inch thick two pounds of Jerusalem artichokes. Fry them in a little butter until brown; fry also brown half-a-pound of sliced onions. Put these to boil in two quarts of water with two turnips, a carrot sliced, two teaspoonfuls of salt, and one of pepper. When the vegetables are tender drain the liquor, set it aside to cool, and remove all fat. Pass the vegetables through a fine sieve to a nice smooth *purée*. Those who possess a Kent's "triturating strainer" will be able to do this much more satisfactorily, both as regards time and results, than by the old way of rubbing through a sieve. Put the liquor on to boil, dissolve in it — according to the strength the soup is required to be — the contents of one or two tins of Nelson's Extract of Meat, then add the vegetable *purée*, a lump or two of sugar, and if required, salt and pepper. Let it boil up and serve.

TURTLE SOUP.

This soup is so often required for invalids, as well as for the table, that an easy and comparatively inexpensive method of preparing it cannot fail to be acceptable. Nelson's Beef Tea or Extract of Meat will be used instead of fresh beef, and Bellis's Sun-dried [20] Turtle instead of live turtle. If convenient it is desirable to soak the dried turtle all night, but it can be used without doing so. Put it on to boil in the water in which it was soaked, in the proportion of one quart with a teaspoonful of salt to a quarter of a pound of the turtle. Add two or three onions peeled and quartered, a small bit of mace and sliced lemon-peel, and simmer gently for four or five hours, or until the turtle is tender enough to divide easily with a spoon. Stock of any kind may be used instead of water, and as the liquid boils away more should be added, to keep the original quantity. Herbs for the proper flavouring of the Turtle Soup are supplied by Bellis; these should be put in about an hour before the turtle is finished, and be tied in muslin. When done take out the turtle and divide it into neat little pieces; strain the liquor in which it was cooked, and having boiled it up, stir in the contents of two tins of Nelson's Extract of Meat, previously soaked for a few minutes. Mix smooth in a gill of cold water a teaspoonful of French potato-flour and of Vienna flour, stir into the soup, and when it has thickened put in the turtle meat; let it get hot through, add a wine-glassful of sherry, a dessertspoonful of lemon-juice, and salt and pepper to taste, and serve at once. It is necessary to have "Bellis's Sun-dried Turtle," imported by T. K. Bellis, Jeffrey's Square, St. Mary Axe, London (sold in boxes), for this soup, because it is warranted properly prepared. An inferior article, got up by negroes from turtle found dead, is frequently sold at a low price; but it is unnecessary to say it is not good or wholesome.

[21]

MOCK TURTLE SOUP.

This, like real turtle soup, can be made of Nelson's Extract of Meat and Bellis's Mock Turtle Meat. Boil the contents of a tin of this meat in water or stock, salted and flavoured with vegetables and turtle herbs, until tender. Finish with Nelson's Extract of Meat, and as directed for turtle soup.

GRAVY.

For roast meat, merely dissolve, after a little soaking, a tin of Nelson's Extract of Meat in a pint of boiling water. For poultry or game, fry two onions a light brown, mince a little carrot and turnip, put in half a teaspoonful of herbs, tied in muslin, and boil until tender, in a pint of water. Strain out the herbs, let the liquor boil up, stir in the contents of a tin of Nelson's Extract of Meat, and if the gravy is required to be slightly thickened, add a small teaspoonful of potato-flour mixed smooth in cold water. For cutlets or other dishes requiring sharp sauce, make exactly as above, and just before serving add a little of any good piquant sauce, or pickles minced finely.

GLAZE.

Soak in a small jar the contents of a tin of Nelson's Extract of Meat in rather less than a gill of cold water. Set the jar over the fire in a saucepan with boiling water, and let the extract simmer until dissolved. This is useful for strengthening soups and gravies, and for glazing ham, tongues, and other things.

[22]

LITTLE DISHES OF FISH.

The recipes we are now giving are suitable for dinner, supper, or breakfast dishes, and will be found especially useful for the latter meal, as there is nothing more desirable for breakfast than fish. We are constantly told that it is not possible to have fresh fish for breakfast, because it cannot be kept all night in the home larder. But we must insist that there is no greater difficulty in keeping fish than meat. Indeed, there is perhaps less difficulty, because fish can be left lying in vinegar, if necessary, whereas in the case of meat it cannot always be done.

We will suppose that it is necessary to use strict economy. It is as well to proceed on that supposition, because people can always be lavish in their expenditure, whereas it is not so easy to provide for the household at once well and economically. In many neighbourhoods fish is sold much cheaper late in the day than in the morning, and in this case the housekeeper who can buy overnight for the use of the next day has a great advantage. Suppose you get the tail of a cod weighing three pounds, as you frequently may, at a very small price in the evening, and use a part of it stuffed and baked for supper, you can have a dish of cutlets of the remainder for breakfast which will be very acceptable. We do not mean a dish of the cold remains, but of a portion of the fish kept uncooked, as it easily may be, as we have before said, by dipping it in vinegar. Or, you get mackerel. Nothing is better than this fish treated according to the recipe we give. [23] Even so delicate a fish as whiting may, by a little management with vinegar, be kept perfectly well from one day to the other. Skinned whiting has very little flavour, and although when skilfully cooked in the usual way it is useful by way of change, the nourishment is much impaired by the removal of the skin. The same remark applies to soles. By frying fish unskinned you get a dish of a different character to that of skinned fish, and one of which the appetite does not so soon tire.

FRIED SOLE.

Soles weighing from three-quarters of a pound to a pound are the most suitable size for frying whole. If it is desired to have the fish juicy and with their full flavour, do not have them skinned. The black side of the soles will not of course look so well, or be so crisp, as the white side, but this is of little consequence compared to the nourishment sacrificed in removing the skin. Have the soles scraped, wipe them, put a tablespoonful of vinegar in a dish, pass the fish through it, and let them lie an hour or more, if necessary all night, as the flavour is thus improved. Run a knife along the backbone, which prevents it looking red when cut. When ready to crumb the fish, lay them in a cloth and thoroughly dry them. Beat up the yolk of an egg with a very little of the white, which will be sufficient to egg a pair of soles; pass the fish through the egg on both sides, hold it up to drain; have ready on a plate a quarter of a pound of very fine dry crumbs, mixed with two ounces of flour, a teaspoonful of salt, and half a teaspoonful of pepper. Draw the [24] fish over the crumbs, first on one side, then on the other, and lay it gently on a dish, black side downwards, whilst you prepare another. Some people succeed better in crumbing fish by sifting the crumbs on to it through a very fine strainer after it is egged. When the fish are ready put them, black side downwards, into the frying-pan with plenty of fat, hot enough to brown a piece of bread instantaneously, move the pan about gently, and when the soles have been fried four minutes, put a strong cooking-fork into them near the head, turn the white side downwards, and fry three minutes longer. Seven minutes will be sufficient to fry a sole weighing three-quarters of a pound, and a pair of this weight is sufficient for a party of six persons. When the sole is done put the fork into the fish close to the head, hold it up and let all the fat drain away, lay it on a sheet of cap paper, and cover over with another sheet. Being thus quite freed from grease, of a rich golden brown, crisp, and with an even surface, lay the fish on the dish for serving, which should have on it either a fish-paper or a napkin neatly folded. A well-fried sole is best eaten without any sauce, but in deference to the national usage, butter sauce, or melted butter, may be served with it.

FILLETED SOLES.

It is better for the cook to fillet the soles, for there is often much waste when it is done by the fishmonger. Having skinned the fish, with a sharp knife make an incision down the spine-bone from the head to the tail, and then along the fins; press the knife [25] between the flesh and the bone, bearing rather hard against the latter, and the fillets will then be readily removed. These can now be dressed in a variety of ways; perhaps the most delicate for breakfast is the following:

FILLETS OF SOLE SAUTÉS.

Having dried the fillets, divide them into neat pieces two or three inches long; dip them in the beaten yolk of egg, and then in seasoned bread-crumbs. Make a little butter hot in the frying-pan, put in the fillets and cook them slowly until brown on one side, then turn and finish on the other.

FILLETS OF SOLE FRIED.

These may either be rolled in one piece or divided into several, as in the foregoing recipe. In either case egg and crumb them thoroughly, place them in the wire-basket as you do them, which immerse in fat hot enough to crisp bread instantly. When done, put the fillets on paper to absorb any grease clinging to them, and serve as hot as possible. All kinds of flat fish can be filleted and cooked by these recipes, and will usually be found more economical than serving the fish whole. It is also economical to fillet the tail-end of cod, salmon, and turbot, and either fry or *sauté*, as may be preferred.

FILLETS OF SOLE WITH LOBSTER.

Thin and fillet a pair of soles, each weighing about a pound. Roll the fillets, secure them with thread, which remove before serving; put them in a stewpan with two ounces of sweet butter, cover closely, and [26] allow them to cook at a slow heat for twenty minutes or until tender, taking care to keep them from getting brown. Prepare a sauce by boiling a quarter of a pound of veal cutlet and the bones of the fish in half-a-pint of water. When reduced to a gill, strain and take off all fat from the sauce, thicken either with fine flour or "Rizine," put it into the stewpan with the fish, and allow it to stand for a quarter of an hour without boiling. Mince or cut in small pieces either the meat of a small fresh lobster, or half a flat tin of the best brand of preserved lobster. Make this hot by putting it in a jam pot standing in a saucepan of boiling water. Take up the fish, carefully pour the sauce round, and place on the top of each fillet some of the lobster.

BAKED WHITING.

Small whiting answer well for this purpose. Tie them round, the tail to the mouth, dip them in dissolved butter, lightly sprinkle with pepper and salt, strew them with pale raspings, put them in a baking-dish with a little butter, and bake in a quick oven for a quarter of an hour.

COD CUTLETS.

A cheap and excellent dish is made by filleting the tail of cod, egging and crumbing the pieces and frying them. Get about a pound and a half of the tail of a fine cod; with a sharp knife divide the flesh from the bone lengthways, cut it into neat pieces as nearly of a size as you can, and flatten with a knife. Dip in egg, then in crumbs mixed with a little flour, pepper, and salt. It is best to fry the cutlets in the wire-basket in [27] plenty of fat, but if this is not convenient they can be done in the frying-pan; in any case, they should be done quickly, so that they may get crisp.

FRIED HERRINGS.

Take care the fish is well cleaned, without being split. Two or three hours before cooking, lightly sprinkle with salt and pepper; when ready to cook, wipe and flour the herrings. Have ready in the frying-pan as much fat at the proper temperature as will cover the herrings. Cook quickly at first, then moderate the heat slightly, and fry for ten to twelve minutes, when they should be crisp and brown. When done, lay them on a dish before the fire, in order that all fat and the fish-oil may drain from them; with this precaution, fried herrings will be found more digestible than otherwise they would be.

ROLLED HERRINGS.

Choose the herrings with soft roes. Having scraped and washed them, cut off the heads, split open, take out the roes, and cleanse the fish. Hold one in the left hand, and, with thumb and finger of the right, press the backbone to loosen it, then lay flat on the board and draw out the bone; it will come out whole, leaving none behind. Dissolve a little fresh butter, pass the inner side of the fish through it, sprinkle pepper and salt lightly over, then roll it up tightly with the fin and tail outwards, roll it in flour and sprinkle a little pepper and salt, then put a small game skewer to keep the herring in shape. Have ready a good quantity of boiling fat; it is best to do the herrings in a wire-basket, and fry them quickly for ten minutes. Take them up and set them on a plate [28] before the fire, in order that all the fat may drain from them. Pass the roes through flour mixed with a sufficient quantity of pepper and salt, fry them brown, and garnish the fish with them and crisp parsley. A difficulty is often felt in introducing herrings at dinner on account of the number of small bones in them, but this is obviated by the above method of dressing, as with care not one bone should be left in.

GALANTINE OF FISH.

Procure a fine large fresh haddock and two smaller, of which to make forcemeat. Take off the head and open the large fish. Carefully press the meat from the backbone, which must be removed without breaking the skin; trim away the rough parts and small bones at the sides. Cover the inside of the fish with a layer of forcemeat, and at intervals place lengthways a few fillets of anchovies, between which sprinkle a little lobster coral which has been passed through a wire sieve; fold the haddock into its original form, and sew it up with a needle and strong thread. Dip a cloth in hot water, wring it as dry as possible, butter sufficient space to cover the fish, then fold it up, tie each end, and put a small safety pin in the middle to keep it firm. Braise the galantine for an hour in stock made from the bones of the fish. Let it stay in the liquor until cold, when take it up and draw out the sewing thread. Reduce and strain the liquor, mix with cream and aspic jelly, or Nelson's Gelatine, dissolved in the proportion of half-an-ounce to a pint. When this sauce is on the point of setting, coat the galantine with it, sprinkle with little passed lobster coral, dish in a bed of shred salad, tastefully interspersed with beetroot cut in dice and dipped in oil and vinegar.

[29] To make the forcemeat, pound the fillets of the small haddocks till fine, then work in about half its quantity of bread panada, an ounce of butter, and the fillets of two anchovies; season with salt and pepper, mix in one egg and a yolk, pass through a wire sieve, and work into it a gill of cream.

FILLETS OF SOLE EN ASPIC.

Aspic jelly, or meat jelly, may be made very good, and at a moderate cost, by boiling lean beef or veal in water with a little vegetable and spice. To make it according to the standard recipes is so expensive and tedious that few persons care to attempt it. The following directions will enable a cook to make an excellent and clear aspic.

Cut two pounds of lean beefsteak or veal cutlet into dice, put it on in two quarts of cold water, and as soon as it boils, take off the scum as it rises. Let it simmer gently for half-an-hour; then add four onions, a turnip, carrot, small bundle of sweet herbs, blade of mace, half-a-dozen white peppercorns, and when it has again boiled for an hour strain it through a napkin. Let it stand until cold, remove all the fat, boil it up, and to a quart of the liquor put an ounce of Nelson's Gelatine, previously soaked in cold water. Add salt and a pinch of cayenne pepper, and when the jelly is cool stir in the whites and shells of two eggs well beaten. Let the jelly boil briskly for two minutes, let it stand off the fire for a few minutes, then strain through a jelly-bag and use as directed. Take the fillets of a pair of large thick soles, cut them into neat square pieces, leaving the trimmings for other dishes, and lay them in vinegar with a little salt for an hour. As they must be kept very white the best French [30] vinegar should be used. Boil the fillets gently in salted water, with a little vinegar, till done; take them up and dry them on a cloth. Have ready some picked parsley and hard-boiled eggs cut in quarters; arrange these neatly at the bottom of a plain mould so as to form a pretty pattern. Pour in very gently enough jelly to cover the first layer, let it stand until beginning to set, then put another layer of fish, eggs, and parsley, then more jelly, and so on until the mould is full. When done set the mould on ice, or allow it to stand some hours in a cold place to get well set. Turn it out, ornament with parsley, beetroot, and cut lemon.

COLLARED EELS.

Clean and boil the eels in water highly seasoned with pepper and salt, an onion, bay-leaf, a clove, and a little vinegar. When the eels are done enough, slip out the bones and cut them up into pieces about two inches long. Take the liquor in which the fish is boiled, strain it, let it boil in the stewpan without the lid, skimming it until it becomes clear. Dissolve a quarter of an ounce of Nelson's Gelatine to each half-pint of the fish gravy, and boil together for a minute, let it then stand until cool. Arrange the pieces of eel tastefully in a plain mould with small sprigs of curled parsley and slices of hard-boiled eggs, and, if you like, a fillet or two of anchovies cut up into dice. When all the fish is thus arranged in the mould, pour the jelly in very gently, a tablespoonful at a time, in order not to disturb the solid material. Let the mould stand in cold water for seven or eight hours, when it can be turned out. Ornament with parsley, lemon, and beetroot.

[31]

LITTLE DISHES OF MEAT.

In this chapter a number of useful and inexpensive dishes are given, which will serve either as breakfast dishes, *entrées*, or for invalids, and which may, in the hands of an intelligent cook, serve as models for many others. As will be seen, it is not so much a question of expense to provide these little tasty dishes as of management. In all the following recipes for little dishes of mutton, it will be found a great advantage to use New Zealand Meat.

A good cook will never be embarrassed by having too much cold meat on hand, because she will be able by her skill so to vary the dishes that the appetites of those for whom she caters will never tire of it. Even a small piece of the loin of mutton may be served in half-a-dozen different ways, and be relished by those who are tired of the mutton-chop or the plain roast.

MUTTON CUTLETS.

Taken from the neck, mutton cutlets are expensive, but those from the loin will be found not only convenient, but to answer well at a smaller cost.

First remove the under-cut or fillet from about two pounds of the best end of a loin of mutton, cut off the flap, which will be useful for stewing, and it is especially good eaten cold, and then remove the meat from the bones in one piece, which divide with the fillet into cutlets about half-an-inch thick. Egg them over and dip them in well-seasoned bread-crumbs, fry them until a nice brown, and serve with gravy made from the bones and an onion.

This way of cooking the loin is much more economical than in chops, because with them the bones and flap are wasted, whereas in cutlets all is used up.

To stew the flap, put it in a stewpan, the fat downwards, sprinkle pepper and salt, and slice an onion or two over, and set it to fry gently in its own fat for an hour. Take up the meat, and put half-a-pint of cold water to the fat, which, when it has risen in a solid cake, take off, mix a little flour with the gravy which will be found beneath the fat, add pepper, salt, and some cooked potatoes cut in slices. Cut the meat into neat squares; let it simmer gently in the gravy with the potatoes for an hour.

ROULADES OF MUTTON.

Remove the fillet from a fine loin of mutton, trim away every particle of skin, fat, and gristle. Flatten the fillet with a cutlet-bat, and cut it lengthways into slices as thin as possible; divide these into neat pieces about three inches long. Sprinkle each with pepper, salt, and finely-chopped parsley, roll them up tightly, then dip in beaten egg, and afterwards in finely-sifted bread-crumbs mixed with an equal quantity of flour and highly seasoned with pepper and salt. As each roulade is thus prepared place it on a game-skewer, three or four on each skewer. [33] Dissolve an ounce of butter in a small frying-pan, and cook the roulades in it.

MUTTON COLLOPS.

Cut neat thin slices from a leg of either roasted or boiled mutton, dip them in yolk of egg and in fine dry bread-crumbs to which a little flour, pepper, and salt have been added. Heat enough butter in a small frying-pan to just cover the bottom, put in the slices of mutton and cook them very slowly, first on one side then on the other, until they are brown. Garnish the dish on which the mutton is served with some fried potatoes or potato chips.

MUTTON SAUTÉ.

Put a little butter or bacon fat in the frying-pan, sprinkle pepper and salt over slices of cold mutton, and let them get hot very slowly. The mutton must be frequently turned, and never allowed to fry. When turned in the pan for the last time sprinkle a little chopped parsley on the upper side; remove the slices carefully on to a hot dish, pour the fat in the pan over, and serve.

COLD MUTTON POTTED.

Cut up the mutton, being careful to free it from all sinew and skin; chop or pound it with half its weight of cooked bacon until it is as fine as desired. Season with a little pepper, salt, and allspice, put it into a jar, which set in a saucepan of water over the fire until the meat is hot through. When taken up stir [34] occasionally until cool, then press it into little pots, and pour clarified butter or mutton fat over the top. If liked, a little essence of anchovy may be added to the seasoning.

MUTTON PIES.

Mince a quarter of a pound of underdone mutton, taking care to have it free from skin and fat. Mix with it a tablespoonful of rich gravy — that which is found under a cake of dripping from a joint is particularly suitable for this purpose — add a few drops of essence of anchovy, a pinch of cayenne pepper, and a small teaspoonful of minced parsley. If necessary add salt.

Line four patty-pans with puff paste, divide the mutton into equal portions and put it into the pans, cover each with a lid of paste, and bake in a quick oven for half-an-hour.

OX BRAIN.

Having carefully washed the brain, boil it very fast, in order to harden it, in well-seasoned gravy. When it is done, take it out of the gravy and set it aside until cold. Cut it either in slices or in halves, dip each piece in egg, then in bread-crumbs well seasoned with dried and sifted parsley, pepper, and salt, fry them in a little butter until brown. The gravy having become cold, take off the fat, and boil it in a stewpan without a lid until it is reduced to a small quantity; pour it round the brain, and serve.

[35]

BRAIN FRITTERS.

Carefully wash an ox brain, and boil it for a quarter of an hour in well-seasoned stock. When the brain is cold, cut it into slices as thin as possible, dip each of them in batter, drop them as you do them into a stewpan half-full of fat at a temperature of 430°, or that which will brown instantly a piece of bread dipped into it. To make the batter, mix two large tablespoonfuls of fine flour with four of cold water, stir in a tablespoonful of dissolved butter or of fine oil, the yolk of an egg, and a pinch of salt and pepper; when ready to use, beat the white of the egg to a strong froth, and mix with it. Do not fry more than two fritters at once; as you take them up, throw them on paper to absorb any grease clinging to them, serve on a napkin or ornamental dish-paper. If this recipe is closely followed, the fritters will be light, crisp, delicate morsels, melting in the mouth, and form besides a very pretty dish. Garnish with fried parsley; take care the parsley is thoroughly dry, put it into a small frying-basket, and immerse it for an instant in the fat in which the fritters are to be cooked. Turn it out on paper, dry, and serve.

MARROW TOAST.

Let the butcher break up a marrow-bone. Take out the marrow in as large pieces as possible, and put them into a stewpan with a little boiling water, rather highly salted. When the marrow has boiled for a [36] minute, drain the water away through a fine strainer. Have ready a slice of lightly-toasted bread, place the marrow on it, and put it into a Dutch oven before the fire for five minutes, or until it is done. Sprinkle over it a little pepper and salt, and a small teaspoonful of parsley, chopped fine. The toast must be served very hot.

CHICKEN IN ASPIC JELLY.

Cut the white part of a cold boiled chicken, and as many similar pieces of cold ham, into neat rounds, not larger than a florin. Run a little aspic jelly into a fancy border mould, allow it to set, and arrange a decoration of boiled carrot and white savoury custard cut crescent shape, dipping each piece in melted aspic. Pour in a very little more jelly, and when it is set place the chicken and ham round alternately, with a sprig of chervil, or small salad, here and there. Put in a very small quantity of aspic to keep this in place, then, when nearly set, sufficient to cover it. Arrange another layer, this time first of ham then of chicken, fix them in the same way, and fill up the mould with aspic jelly. When the dish is turned out fill the centre with cold green peas, nicely seasoned, and garnish round with chopped aspic and little stars of savoury custard. To make this, soak a quarter of an ounce of Nelson's Gelatine in a gill of milk, dissolve it over the fire, and stir in a gill of thick cream, season to taste with cayenne pepper and salt, and, if liked, a little grate of nutmeg. Pour the [37] custard on to a large dish, and when cold cut it into the required shapes.

VEAL CUTLETS IN WHITE SAUCE.

Cut six or seven cutlets, about half-an-inch thick, from a neck of veal, braise them in half-a-pint of good white stock with an onion, a small bunch of herbs, a bacon bone, and two or three peppercorns, until they are done. Let the cutlets get cool in the liquor, then drain them. Strain the liquor and make a white sauce with it; add a tablespoonful of thick cream and a quarter of an ounce of Nelson's Gelatine, dissolved in a gill of milk; season with salt and cayenne pepper, stirring occasionally until quite cold. Dip the cutlets in, smoothly coating one side, and before the sauce sets decorate them with very narrow strips of truffle in the form of a star. Cut as many pieces of cooked tongue or ham as there are cutlets, dish them alternately in a circle on a border of aspic, fill the centre with a salad composed of all kinds of cold cooked vegetables, cut with a pea-shaped cutter and seasoned with oil, vinegar, pepper, and salt. Garnish with aspic jelly cut lozenge shape and sprigs of chervil.

KIDNEYS SAUTÉS.

Like many other articles of diet, kidneys within the last ten years have been doubled in price, and are so scarce as to be regarded as luxuries. The method of cooking them generally in use is extravagant, and renders them tasteless and indigestible. Kidneys [38] should never be cooked rapidly, and those persons who cannot eat them slightly underdone should forego them. One kidney dressed as directed in the following recipe will go as far as two cooked in the ordinary manner—an instance, if one were needed, of the economy of well-prepared food.

Choose fine large kidneys, skin them and cut each the round way into thin slices: each kidney should yield from ten to twelve slices. Have ready a tablespoonful of flour highly seasoned with pepper and salt and well mixed together; dip each piece of kidney in it. Cut some neat thin squares of streaked bacon, fry them *very slowly* in a little butter; when done, put them on the dish for serving, and keep hot whilst you *sauté* the kidneys, which put into the fat the bacon was cooked in. In about a minute the gravy will begin to rise on the upper side, then turn the kidneys and let them finish cooking slowly; when they are done, as they will be in three to four minutes, the gravy will again begin to rise on the side which is uppermost. Put the kidneys on the dish with the bacon, and pour over them a spoonful or two of plain beef gravy, or water thickened with a little flour, boiled and mixed with the fat and gravy from the kidneys in the frying-pan. If there is too much fat in the pan, pour it away before boiling up the gravy. Serve the kidneys on a hot-water dish.

TINNED KIDNEYS WITH MUSHROOMS.
(*Tomoana Brand.*)

Dry a half-tin of champignons in a cloth, or, if convenient, prepare a similar quantity of fresh button [39] mushrooms; add to these a few pieces of dried mushrooms, previously soaked for ten minutes in tepid water, put them into a stewpan with a slice of butter, and stir constantly for six minutes, then add two or three kidneys cut in small neat pieces, in the shape of dice is best, and continue stirring until the kidneys are hot through, taking care to do them slowly; at the last moment season with pepper and salt, and serve very hot. Garnish the dish with fried sippets of bread.

KIDNEYS WITH PICCALILLI SAUCE.
(*Tomoana Brand.*)

Take the kidneys out of the gravy, and cut them into six slices. Mix a small teaspoonful of curry powder with three teaspoonfuls of fine flour and a small pinch of salt. Dip each slice in this mixture, and when all are done put them in the frying-pan with a little butter, and let them get slowly hot through. When done, put the kidneys in the centre of a hot dish, and pour round them a sauce made as follows: Boil up the gravy of the kidneys, and stir into it sufficient minced piccalilli pickles to make it quite thick, add a teaspoonful of flour to a tablespoonful of the piccalilli vinegar, stir into the sauce, and when all has boiled up together, pour it round the kidneys.

BROILED KIDNEYS.

These are quite an epicure's dish, and care must be taken to cook them slowly. Having skinned the kidneys (they must not be split or cut) dip them for a [40] moment in boiling fat, place them on the gridiron over a slow fire, turning them every minute. They will take ten to fifteen minutes to cook, and will be done as soon as the gravy begins to run. Place them on a hot dish rubbed over with butter, salt and pepper them rather highly. It must be understood that kidneys thus cooked ought to have the gravy in them, and that when they are cut at table it should run from them freely and in abundance.

LAMB'S FRY.

A really proper fry should consist not only of sweetbreads and liver, but of the heart, melt, brains, frill, and kidneys, each of which requires a different treatment. It is quite as easy to cook a fry properly as to flour and fry it hard and over-brown, as is too frequently done. Trim the sweetbreads neatly, and simmer them for a quarter of an hour in good white stock with an onion. When they are done take them up and put the brains in the gravy, allowing them to boil as fast as possible in order to harden them; let them get cold, then cut into slices, egg and bread-crumb them, and fry with the sweetbread in a little butter. After the brains are taken out of the gravy, put the slices of heart and melt in, and let them stew slowly until tender. When they are ready, flour them, and fry with the liver and frill until brown. Lastly, put the kidneys, cut in slices, into the pan, and very gently fry for about a minute. Shake a little flour onto the pan, stir it about until it begins to brown; then pour on to it the gravy, in which the sweetbreads, etc., [41] were stewed, see it is nicely seasoned, and pour round the fry, which should be neatly arranged in the centre of the dish. Garnish with fried parsley.

LAMB'S SWEETBREADS.

These make an admirable breakfast dish, and can be partly prepared over-night. Trim and wash the sweetbreads, put them into a saucepan with sufficient well-flavoured stock to cover them, a minced onion and a sprig of lemon-thyme; boil gently for fifteen minutes, or a little longer if necessary. Take them up, drain, dip in egg and finely-sifted bread-crumbs mixed with a little flour, pepper, and salt. Fry very carefully, so as not to make it brown or hard, some small slices of bacon, keep warm whilst you fry the sweetbreads in the fat which has run from it, adding, if required, a little piece of butter or lard. For a breakfast dish, the sweetbreads should be served without gravy, but if for an *entrée* the liquor in which they were stewed, with slight additions and a little thickening, can be poured round them in the dish. Calves' sweetbreads are prepared in the same manner as the above, and can either be fried, finished in a Dutch oven, or served white, with parsley and butter, or white sauce.

VEAL À LA CASSEROLE.

For this dish a piece of the fillet about three inches thick will be required, and weighing from two to three pounds. It should be cut from one side of the leg, [42] without bone; but sometimes butchers object to give it, as cutting in this manner interferes with cutlets. In such a case a piece must be chosen near the knuckle, and the bone be taken out before cooking. For a larger party, a thick slice of the fillet, weighing about four pounds, will be found advantageous.

With a piece of tape tie the veal into a round shape, flour, and put it into a stewpan with a small piece of butter, fry until it becomes brown on all sides. Then put half a pint of good gravy, nicely seasoned with pepper and salt, cover the stewpan closely, and set it on the stove to cook very slowly for at least four hours. When done, the veal will be exquisitely tender, full of flavour, but not the least ragged. Take the meat up, and keep hot whilst the gravy is reduced, by boiling without the lid of the saucepan, to a rich glaze, which pour over the meat and serve.

BROWN FRICASSÉE OF CHICKEN.

This is a brown fricassée of chicken, and is an excellent dish. No doubt the reason it is so seldom given is that, although easy enough to do, it requires care and attention in finishing it. Many of the best cooks, in the preparation of chickens for fricassée, cut them up before cooking, but we prefer to boil them whole, and afterwards to divide them, as the flesh thus is less apt to shrink and get dry. The chicken can be slowly boiled in plain water, with salt and onions, or, as is much better, in white broth of any kind. When the chicken is tender cut it up; take [43] the back, and the skin, pinions of the wings, and pieces which do not seem nice enough for a superior dish, and boil them in a quart of the liquor in which it was boiled. Add mushroom trimmings, onions, and a sprig of thyme; boil down to one-half, then strain, take off all fat, and stir over the fire with the yolk of two eggs and an ounce of fine flour until thickened. Dip each piece of chicken in some of this sauce, and when they are cold pass them through fine bread-crumbs, then in the yolk of egg, and crumb again. Fry carefully in hot fat. Dish the chicken with a border of fried parsley, and the remainder of the gravy poured round the dish. This dish is generally prepared by French cooks by frying the chicken in oil, and seasoning with garlic; but unless the taste of the guests is well known, it is safer to follow the above recipe.

CHICKEN SAUTÉ.

Put any of the meat of the breast or of the wings without bone into a frying-pan with a little fresh butter or bacon fat. Cook them very slowly, turning repeatedly; if the meat has not been previously cooked it will take ten minutes, and five minutes if a *réchauffé*. Sprinkle with pepper, and serve with mushrooms or broiled bacon. The legs of cooked chickens are excellent *sautés*, but they should be boned before they are put into the pan.

POTATO HASH.

Put some cold potatoes chopped into the frying-pan with a little fat, stir them about for five minutes, [44] then add to them an equal quantity of cold meat, cut into neat little squares, season nicely with pepper and salt, fry gently, stirring all the time, until thoroughly hot through.

DRY CURRY.

Fry a minced onion in butter until lightly browned, cut up the flesh of two cooked chicken legs, or any other tender meat, into dice, mix this with the onions, and stir them together over the fire until the meat is hot through; sprinkle over it about a small teaspoonful of curry-powder, and salt to taste. Having thoroughly mixed the meat with the curry-powder, pour over it a tablespoonful of milk or cream, and stir over the fire until the moisture has dried up. Celery salt may be used instead of plain salt, and some persons add a few drops of lemon-juice when the curry is finished.

CROQUETTES.

Croquettes of all kinds, fish, game, poultry or any delicate meats, can be successfully made on the following model: Whatever material is used must be finely minced or pounded. Care is required in making the sauce, if it is too thin it is difficult to mould the croquettes, and ice will be required to set it. Croquettes of game without any flavouring, except a little salt and cayenne, are generally acceptable as a breakfast dish. Preserved lobster makes very good croquettes for an *entrée*, and small scraps of any kind can thus be made into a very good dish. Put one ounce of fine flour into a stewpan with half a gill of cold [45] water, stir this over a slow fire very rapidly until it forms a paste, then add one ounce of butter, and stir until well incorporated. Mix in a small teaspoonful of essence of shrimps or anchovies, with a pinch of salt and pepper. Take the stewpan off the fire, and stir the yolk of an egg briskly into the sauce; thoroughly mix it with half-a-pound of pounded fish or meat, spread it out on a plate until it is cool. Flour your hands, take a small piece of the croquette mixture, roll into a ball or into the shape of a cork, then pass it through very finely-sifted and dried bread-crumbs. Repeat the process until all the mixture is used; put the croquettes as you do them into a wire frying-basket, which shake very gently, when all are placed in it, in order to free them from superfluous crumbs. Have ready a stewpan half-full of boiling fat, dip the basket in, gently moving it about, and taking care the croquettes are covered with fat. In about a minute they will become a delicate brown, and will then be done. Turn them on a paper to absorb any superfluous fat, serve them on a napkin or ornamental dish paper. No more croquettes than will lie on the bottom of the basket without touching each other should be fried at once.

MEAT CAKES À L'ITALIENNE.

Mix very fine any kind of cold meat or chicken, taking care to have it free from skin and gristle, add to it a quarter of its weight of sifted bread-crumbs, a few drops of essence of anchovy, a little parsley, pepper and salt, and sufficient egg to moisten the [46] whole. Flour your hands, roll the meat into little cakes about the size of a half-crown piece, then flatten the cakes with the back of a spoon, dip them in egg and fine bread-crumbs, and fry them in a little butter until lightly browned on the outside. Put them on a hot dish and garnish with boiled Italian paste.

RAISED PORK PIE.

Take a pound of meat, fat and lean, from the chump end of a fine fore-loin of pork, cut it into neat dice, mix a tablespoonful of water with it, and season with a large teaspoonful of salt and a small one of black pepper. To make the crust, boil a quarter of a pound of lard or clarified dripping in a gill and a half of water, and pour it hot on to one pound of flour, to which a good pinch of salt has been added. Mix into a stiff paste, pinch off enough of it to make the lid, and keep it hot. Flour your board and work the paste into a ball, then with the knuckles of your right hand press a hole in the centre, and mould the paste into a round or oval shape, taking care to keep it a proper thickness. Having put in the meat, join the lid to the pie, which raise lightly with both hands so as to keep it a good high shape, cut round the edge with a sharp knife, and make the trimmings into leaves to ornament the lid; and having placed these on, with a rose in the centre, put the pie on a floured baking-sheet and brush it over with yolk of egg.

The crust of the pie should be cool and set before putting it into the oven, which should be a moderate [47] heat. When the gravy boils out the pie is done. An hour and a half will bake a pie of this size. Make a little gravy with the bones and trimmings of the pork, and to half-a-pint of it add a quarter of an ounce of Nelson's Gelatine, and nicely season with pepper and salt. When the pie is cold remove the rose from the top, make a little hole, insert a small funnel, and pour in as much gravy as the pie will hold. Replace the rose on the top, and put the pie on a dish with a cut paper.

If preferred, the pie can be made in a tin mould; but the crust is nicer raised by the hand. A great point to observe is to begin moulding the crust whilst it is hot, and to get it finished as quickly as possible.

VEAL AND HAM PIE.

Prepare the crust as for a pork pie. Cut a pound of veal cutlet and a quarter of a pound of ham into dice, season with a teaspoonful of salt and another of black pepper, put the meat into the crust, and finish as for pork pie. Add a quarter of an ounce of Nelson's Gelatine—previously soaked in cold water, and then dissolved—to a teacupful of gravy made from the veal trimmings.

PORK SAUSAGES.

When a pig is cut up in the country, sausages are usually made of the trimmings; but when the meat has to be bought, the chump-end of a fore-loin will be found to answer best. The fine well-fed meat of a full-grown pig, known in London as "hog-meat," is every way preferable to that called "dairy-fed pork." [48] The fat should be nearly in equal proportion to the lean, but of course this matter must be arranged to suit the taste of those who will eat the sausages. If young pork is used, remove the skin as thinly as you can—it is useful for various purposes—and then with a sharp knife cut all the flesh from the bones, take away all sinew and gristle, and cut the fat and lean into strips. Some mincing-machines require the meat longer than others; for Kent's Combination, cut it into pieces about an inch long and half-an-inch thick. To each pound of meat put half a gill of gravy made from the bones, or water will do; then mix equally with it two ounces of bread-crumbs, a large teaspoonful of salt, a small one of black pepper, dried sage, and a pinch of allspice. This seasoning should be well mixed with the bread, as the meat will then be flavoured properly throughout the mass. Arrange the skin on the filler, tie it at the end, put the meat, a little at a time, into the hopper, turn the handle of the machine briskly, and take care the skin is only lightly filled. When the sausages are made, tie the skin at the other end, pinch them into shape, and then loop them by passing one through another, giving a twist to each as you do them. Sausage-skins, especially if preserved, should be well soaked before using, or they may make the sausages too salt. It is a good plan to put the skin on the water-tap and allow the water to run through it, as thus it will be well washed on the inside. Fifteen to twenty minutes should be allowed for frying sausages, and when done they should be nicely browned. A little butter or lard is best for [49] frying, and some pieces of light bread may be fried in it when the sausages are done, and placed round the dish by way of garnish. Cooks cannot do better than remember Dr. Kitchener's directions for frying sausages. After saying, "They are best when quite fresh made," he adds: "put a bit of butter or dripping into a clean frying-pan; as soon as it is melted, before it gets hot, put in the sausages, and shake the pan for a minute, and keep turning them. Be careful not to break or prick them in so doing. Fry them over a very slow fire till they are nicely browned on all sides. The secret of frying sausages is to let them get hot very gradually; they then will not break if they are not stale. The common practice to prevent them bursting is to prick them with a fork, but this lets the gravy out."

[50]

PUDDINGS.

CUSTARD PUDDING.

We give this pudding first because it affords an opportunity for giving hints on making milk puddings generally, and because, properly made, there is no more delicious pudding than this. It is besides most useful and nutritious, not only for the dinner of healthy people, but for children and invalids. But few cooks, however, make it properly; as a rule too many eggs are used, to which the milk is added cold, and the pudding is baked in a quick oven. The consequence is that the pudding curdles and comes to table swimming in whey; or, even if this does not happen, the custard is full of holes and is tough.

In the first place, milk for all puddings with eggs should be poured on to the eggs boiling hot; in the next, the baking must be very slowly done, if possible, as directed in the recipe; the dish containing the pudding to be placed in another half-full of water. This, of course, prevents the baking proceeding too rapidly, and also prevents the pudding acquiring a sort of burned greasy flavour, which is injurious for invalids. Lastly, too many eggs should not be used; the quantity given, two to the pint of milk, is in all cases quite sufficient, and will make a fine rich custard.

We never knew a pudding curdle, even with [51] London milk a day old, if all these directions were observed; but it is almost needless to say, that the pudding made with new rich milk is much finer than one of inferior milk.

Boil a pint and a half of milk with two ounces of lump sugar, or rather more if a sweet pudding is liked, and pour it boiling hot on three eggs lightly beaten—that is, just sufficiently so to mix whites and yolks. Flavour the custard with nutmeg, grated lemon-peel, or anything which may be preferred and pour it into a tart-dish. Place this dish in another three-parts full of boiling water, and bake slowly for forty minutes, or until the custard is firm. There is no need to butter the dish if the pudding is baked as directed.

SOUFFLÉ PUDDING.

This is a delicious pudding, and to insure its success great care and exactness are required. In the first place, to avoid failure it is necessary that the butter, flour, sugar, and milk, should be stirred long enough over a moderate fire to make a stiff paste, because if this is thin the eggs will separate, and the pudding when done resemble a batter with froth on the top.

Before beginning to make the pudding, prepare a pint tin by buttering it inside and fastening round it with string on the outside a buttered band of writing-paper, which will stand two inches above the tin and prevent the pudding running over as it rises. Melt an ounce of butter in a stewpan, add one ounce of sifted sugar, stir in an ounce and a half of Vienna [52] flour, mix well together, add a gill of milk, and stir over the fire with a wooden spoon until it boils and is thick. Take the stewpan off the fire, beat up the yolks of three eggs with half a teaspoonful of extract of vanilla, and stir a little at a time into the paste, to insure both being thoroughly mixed together. Put a small pinch of salt to the whites of four eggs, whip them as stiff as possible, and stir lightly into the pudding, which pour immediately into the prepared mould. Have ready a saucepan with enough boiling water to reach a little way up the tin, which is best placed on a trivet, so that the water cannot touch the paper band. Let the pudding steam very gently for twenty minutes, or until it is firm in the middle, and will turn out.

For sauce, boil two tablespoonfuls of apricot jam in a gill of water, with two ounces of lump sugar, stir in a wine-glassful of sherry, add a few drops of Nelson's Vanilla Flavouring, pour over the pudding and serve.

OMELET SOUFFLÉ.

Put the yolks of two eggs into a basin with an ounce of sifted sugar and a few drops of Nelson's Vanilla Essence; beat the yolks and sugar together for six minutes, or until the mixture becomes thick. Then whip the whites very stiff, so that they will turn out of the basin like a jelly. Mix the yolks and whites lightly together, have ready an ounce of butter dissolved in the omelet-pan, pour in the eggs, hold this pan over a slow fire for two minutes, then put the frying-pan into a quick oven and bake until the [53] omelet has risen; four minutes ought to be sufficient to finish the omelet in the oven; when done, slide it on to a warm dish, double it, sift sugar over, and serve instantly.

SPONGE SOUFFLÉ.

Cover the bottom of a tart-dish with sponge-cakes, pour over a little brandy and sherry; put in a moderate oven until hot, then pour on the cakes an egg whip made of two packets of Nelson's Albumen, beaten to a strong froth with a little sugar. Bake for a quarter of an hour in a slow oven.

CABINET PUDDING.

Butter very thickly a pint pudding-basin, and cover it neatly with stoned muscatel raisins, the outer side of them being kept to the basin. Lightly fill up the basin with alternate layers of sponge-cake and ratafias, and when ready to steam the pudding, pour by degrees over the cake a custard made of half-a-pint of boiling milk, an egg, three lumps of sugar, a tablespoonful of brandy, and a little lemon flavouring. Cover the basin with a paper cap and steam or boil gently for three-quarters of an hour. Great care should be taken not to boil puddings of this class fast, as it renders them tough and flavourless.

BRANDY SAUCE.

Mix a tablespoonful of fine flour with a gill of cold water, put it into a gill of boiling water, and, having [54] stirred over the fire until it is thick, add the yolk of an egg. Continue stirring for five minutes, and sweeten with two ounces of castor sugar. Mix a wine-glass of brandy with two tablespoonfuls of sherry, stir it into the sauce, and pour it round the pudding. If liked, a grate of nutmeg may be added to the sauce, and, if required to be rich, an ounce of butter may be stirred in before the brandy.

WARWICKSHIRE PUDDING.

Butter a pint-and-a-half tart-dish, lay in it a layer of light bread, cut thin, on this sprinkle a portion of two ounces of shred suet, and of one ounce of lemon candied-peel, chopped very fine. Fill the dish lightly with layers of bread, sprinkling over each a little of the suet and peel.

Boil a pint of milk with two ounces of sugar, pour it on two eggs, beaten for a minute, and add it to the pudding just before putting it into the oven; a little of Nelson's Essence of Lemon or Almonds may be added to the custard. Bake the pudding in a very slow oven for an hour.

VANILLA RUSK PUDDING.

Dissolve, but do not oil, an ounce of butter, mix in a quarter of a pound of sifted sugar, stir over the fire for a few minutes, add an egg well beaten, and half a teaspoonful of Nelson's Vanilla Extract, or as much as will give a good flavour to the paste, which continue stirring until it gets thick. [55]

Spread four slices of rusk with the vanilla paste, put them in a buttered tart-dish. Boil half-a-pint of new milk, pour it on to an egg well beaten, then add it to the rusk, and put the pudding to bake in a slow oven for an hour. Turn out when done, and sift sugar over the pudding. If a superior pudding is desired, boil a tablespoonful of apricot jam in a teacupful of plain sugar syrup, add a little vanilla flavouring, and pour over the pudding at the moment of serving.

JUBILEE PUDDING.

Pour a pint of boiling milk on two ounces of Rizine, stir over the fire for ten minutes, add half an ounce of butter, the yolks of two eggs, an ounce of castor sugar, and six drops of Nelson's Essence of Almonds. Put the pudding into a buttered pie-dish, and bake in a moderate oven for a quarter of an hour. When taken from the oven, spread over it a thin layer of apricot jam, and on this the whites of the eggs beaten to a strong froth, with half an ounce of castor sugar. Return the pudding to a slow oven for about four minutes, in order to set the meringue.

NATAL PUDDING.

Soak half-an-ounce of Nelson's Gelatine in half-a-pint of cold water until it is soft, when add the grated peel of half a lemon, the juice of two lemons, the beaten yolks of three eggs, and six ounces of lump sugar dissolved in half-a-pint of boiling water. Stir the mixture over the fire until it thickens, taking care that it does not boil. Have ready the whites of the eggs [56] well whisked, stir all together, pour into a fancy mould, which put into a cold place until the pudding is set.

QUEEN'S PUDDING.

Half-a-pound of bread-crumbs, a pint of new milk, two ounces of butter, the yolks of four eggs, and a little Nelson's Essence of Lemon. Boil the bread-crumbs and milk together, then add the sugar, butter, and eggs; when these are well mixed, bake in a tart-dish until a light brown. Then put a layer of strawberry jam, and on the top of this the whites of the eggs beaten to a stiff froth, with a little sifted sugar. Smooth over the meringue with a knife dipped in boiling water, and bake for ten minutes in a slow oven.

CHOCOLATE PUDDING.

Boil half-a-pound of light stale bread in a pint of new milk. Stir continually until it becomes a thick paste; then add an ounce of butter, a quarter of a pound of sifted sugar, and two large teaspoonfuls of Schweitzer's Cocoatina, with a little Nelson's Essence of Vanilla. Take the pudding off the fire, and mix in, first, the yolks of three eggs, then the whites beaten to a strong froth. Put into a buttered tart-dish and bake in a moderate oven for three-quarters of an hour.

COCOA-NUT PUDDING.

Choose a large nut, with the milk in it, grate it finely, mix it with an equal weight of finely-sifted sugar, half its weight of butter, the yolks of four eggs, and the milk of the nut. Let the butter be beaten to a cream, [57] and when all the other ingredients are mixed with it, add the whites of the eggs, whisked to a strong froth. Line a tart-dish with puff-paste, put in the pudding mixture and bake slowly for an hour. Butter a sheet of paper and cover the top of the pudding, as it should not get brown.

RASPBERRY AND CURRANT PUDDING.

Stew raspberries and currants with sugar and water, taking care to have plenty of juice. Cut the crumb of a stale tin-loaf in slices about half-an-inch thick and put in a pie-dish, leaving room for the bread to swell, with alternate layers of fruit, until the dish is full. Then put in as much of the juice as you can without causing the bread to rise. When it is soaked up put in the rest of the juice, cover with a plate, and let the pudding stand until the next day. When required for use turn out and pour over it a good custard or cream. The excellence of this pudding depends on there being plenty of syrup to soak the bread thoroughly. This is useful when pastry is objected to.

THE CAPITAL PUDDING.

Shred a quarter of a pound of suet, mix it with half a pound of flour, one small teaspoonful each of baking-powder and carbonate of soda, then add four tablespoonfuls of strawberry or raspberry jam, and stir well with a gill of milk. Boil for four hours in a high mould, and serve with wine or fruit sauce. The latter is made by stirring jam into thin butter sauce.

[58]

ITALIAN FRITTERS.

Cut slices of very light bread half-an-inch thick, with a round paste-cutter, divide them into neat shapes all alike in size. Throw them into boiling fat and fry quickly of a rich golden brown, dry them on paper, place on a dish, and pour over orange or lemon syrup, or any kind of preserve made hot. Honey or golden syrup may be used for those who like them.

DUCHESS OF FIFE'S PUDDING.

Boil two ounces of rice in a pint of milk until quite tender. When done, mix with it a quarter of an ounce of Nelson's Gelatine soaked in a tablespoonful of water. Line the inside of a plain mould with the rice, and when it is set fill it up with half-a-pint of cream, whipped very stiff and mixed with some nice preserve, stewed fruit, or marmalade. After standing some hours turn out the pudding, and pour over it a delicate syrup made of the same fruit as that put inside the rice.

WELSH CHEESECAKE.

Dry a quarter of a pound of fine flour, mix with two ounces of sifted loaf-sugar, and add it by degrees to two ounces of butter beaten to a cream; then work in three well-beaten eggs, flavour with Nelson's Essence of Lemon. Line patty-pans with short crust, put in the above mixture, and bake in a quick oven.

FRIAR'S OMELET.

Make six moderate-sized apples into sauce, sweeten [59] with powdered loaf-sugar, stir in two ounces of butter, and when cold, mix with two well-beaten eggs. Butter a tart-dish, and strew the bottom and sides thickly with bread-crumbs, then put in the apple-sauce, and cover with bread-crumbs to the depth of a quarter of an inch, put a little dissolved butter on the top, and bake for an hour in a good oven. When done, turn it out, and sift sugar over it.

COMPOTE OF APPLES WITH FRIED BREAD.

Bake a dozen good cooking apples, scrape out the pulp, boil this with half-a-pound of sugar to a pound of pulp, until it becomes stiff. It must be stirred all the time it is boiling. When done, place the compote in the centre of the dish, piling it up high. Have ready some triangular pieces of fried bread, arrange some like a crown on the top, the remainder at the bottom of the compote. Have ready warmed half a pot of apricot marmalade mixed with a little plain sugar-syrup, and pour it over the compote, taking care that each piece of bread is well covered.

APPLE FOOL.

Bake good sharp apples; when done, remove the pulp and rub it through a sieve, sweeten and flavour with Nelson's Essence of Lemon; when cold add to it a custard made of eggs and milk, or milk or cream sweetened will be very good. Keep the fool quite thick. Serve with rusks or sponge finger biscuits.

[60]

APPLE MERINGUE.

Beat up two packets of Nelson's Albumen with six small teaspoonfuls of water, and stir them into half-a-pound of stiff applesauce flavoured with Nelson's Essence of Lemon. Put the meringue on a bright tin or silver dish, pile it up high in a rocky shape, and bake in a quick oven for ten minutes.

STEWED PEARS WITH RICE.

Put four large pears cut in halves into a stewpan with a pint of claret, Burgundy, or water, and eight ounces of sugar, simmer them until perfectly tender. Take out the pears and let the syrup boil down to half; flavour it with vanilla. Have ready a teacupful of rice, nicely boiled in milk and sweetened, spread it on a dish, lay the pears on it, pour the syrup over, and serve. This is best eaten cold.

COMPOTE OF PRUNES.

Wash the fruit in warm water, put it on to boil in cold water in which lump sugar has been dissolved. To a pound of prunes put half-a-pound of sugar, a pint of water, with the thin rind and juice of a lemon. Let them simmer for an hour, or until so tender that they will mash when pressed. Strain the fruit and set it aside. Boil the syrup until it becomes very thick and is on the point of returning to sugar, then pour it over the prunes, turn them about so that they become thoroughly coated, taking care not to break them, let them lie for twelve hours, then pile up on a glass dish for dessert.

[61]

ON JELLY-MAKING.

It is within the memory of many persons that jelly was only to be made from calves' feet by a slow, difficult, and expensive process. There is, indeed, a story told of the wife of a lawyer, early in this century, having appropriated some valuable parchment deeds to make jelly, when she could not procure calves' feet. But the secret that it could be so made was carefully guarded by the possessors of it, and it was not until the introduction of Nelson's Gelatine that people were brought to believe that jelly could be made other than in the old-fashioned way. Even now there is a lingering superstition that there is more nourishment in jelly made of calves' feet than that made from Gelatine. The fact is, however, that Gelatine is equally nutritious from whatever source it is procured. Foreign Gelatine, as is well known, does sometimes contain substances which, if not absolutely deleterious, are certainly undesirable; but Messrs. Nelson warrant their Gelatine of equal purity with that derived from calves' feet.

It is unnecessary to enlarge on the economy both in time and money of using Gelatine, or the more certain result obtained from it. If the recipe given for making "a quart of jelly" is closely followed, a most excellent and brilliant jelly will be produced. Many cooks get worried about their jelly-bags, and are [62] much divided in opinion as to the best kind to use. It is not a point of great consequence whether a felt or close flannel is selected. We incline to the latter, which must be of good quality, and if the material is not thick it should be used double.

When put away otherwise than perfectly clean and dry, or when stored in a damp place, flannel bags are sure to acquire a strong mouldy flavour, which is communicated to all jelly afterwards strained through them.

The great matter, therefore, to observe in respect of the jelly-bag, is that it be put away in a proper condition, that is, perfectly free from all stiffness and from any smell whatever.

As soon as the bag is done with, turn it inside out, throw it into a pan of boiling water, stir it about with a spoon until it is cleansed. Then, have another pan of boiling water, and again treat the bag in the same manner. Add as much cold water as will enable you to wring the bag out dry, or it can be wrung out in a cloth. This done, finally rinse in hot water, wring, and, if possible, dry the bag in the open air. See that it is perfectly free from smell; if not, wash in very hot water again. Wrap the bag in several folds of clean paper and keep it in a dry place.

A thing to be observed is that, if the jelly is allowed to come very slowly to boiling-point it will be more effectually cleared, as the impurities of the sugar and the thicker portions of the lemons thus rise more surely with the egg than if this part of the process is too rapidly carried out. In straining, if the jelly is [63] well made, it is best to pour all into the bag at one time, doing it slowly, so as not to break up the scum more than necessary. Should the jelly not be perfectly bright on a first straining, it should be kept hot, and slowly poured again through the bag. The contents of the bag should not be disturbed, nor should the slightest pressure be applied, as this is certain to cloud the jelly. If brandy is used, it should be put in after the jelly is strained, as by boiling both the spirit and flavour of it are lost.

In order that jelly may turn out well, do not put it into the mould until it is on the point of setting. If attention is paid to this there will never be any difficulty in getting jelly to turn out of a mould, and putting it into hot water or using hot cloths will be unnecessary. A mould should be used as cold as possible, because then when the jelly comes into contact with it, it is at once set and cannot stick. Any kind of mould may be used. If the direction to put the jelly in *when just setting* is followed, it will turn out as well from an earthenware as from a copper mould.

It should be unnecessary to say that the utmost cleanliness is imperative to insure the perfection of jelly. So delicate a substance not only contracts any disagreeable flavour, but is rendered cloudy by the least touch of any greasy spoon, or by a stewpan which has not been properly cleansed.

[64]

HOW TO USE GELATINE.

There are a few points connected with the use of Gelatine for culinary purposes which cannot be too strongly impressed upon housekeepers and cooks.

1. Gelatine should always be soaked in cold water till it is thoroughly saturated—say, till it is so soft that it will tear with the fingers—whether this is specified in the recipe or not.

2. Nelson's Gelatine being cut very fine will soak in about an hour, but whenever possible it is desirable to give it a longer time. When convenient, it is a good plan to put Gelatine to soak overnight. It will then dissolve in liquid below boiling-point.

When jelly has to be cleared with white of egg do not boil it longer than necessary. Two minutes is quite sufficient to set the egg and clarify the jelly.

Use as little Gelatine as possible; that is to say, never use more than will suffice to make a jelly strong enough to retain its form when turned out of the mould. The prejudice against Gelatine which existed in former years was doubtless caused by persons unacquainted with its qualities using too large a quantity, and producing a jelly hard, tough, and unpalatable, which compared very unfavourably with the delicate jellies they had been accustomed to make from calves' feet, the delicacy of which arose from the simple fact that the Gelatine derived from calves' feet is so weak that it is almost impossible to make the jellies too strong.

Persons accustomed to use Gelatine will know that its "setting" power is very much affected by the [65] temperature. In the recipes contained in the following pages the quantity of Gelatine named is that which experience has shown to be best suited to the average temperature of this country. In hot weather and foreign climates a little more Gelatine should be added.

TO MAKE A QUART OF BRILLIANT JELLY.

Soak one ounce of Nelson's Opaque Gelatine in half-a-pint of cold water for two or three hours, and then add the same quantity of boiling water; stir until dissolved, and add the juice and peel of two lemons, with wine and sugar sufficient to make the whole quantity one quart; have ready the white and shell of an egg, well beaten together, or a packet of Nelson's Albumen, and stir these briskly into the jelly; boil for two minutes without stirring it; remove from the fire, allow it to stand two minutes, and strain through a close flannel bag. Let it be on the point of setting before putting into the mould.

AN ECONOMICAL JELLY.

For general family use it is not necessary to clear jelly through the bag, and a quart of excellent jelly can be made as follows: Soak one ounce of Nelson's Gelatine in half-a-pint of cold water for two or three hours, then add a 3d. packet of Nelson's Citric Acid and three-quarters of a pound of loaf sugar; pour on half-a-pint of boiling water and half-a-pint of sherry, orange or other wine (cold), and add one-twelfth part of a bottle of Nelson's Essence of [66] Lemon; stir for a few minutes before pouring into the moulds.

The effect of citric acid in the above quantity is to make the jelly clearer. When this is not of consequence, a third of a packet can be used, and six ounces of sugar. Wine can be omitted if desired, and water substituted for it. Ginger-beer makes an excellent jelly for those who do not wish for wine, and hedozone is also very good.

JELLY WITH FRUIT.

This is an elegant sweetmeat, and with clear jelly and care in moulding, can be made by inexperienced persons, particularly if Nelson's Bottled Jelly is used. If the jelly is home-made the recipe for making a "quart of jelly" will be followed. When the jelly is on the point of setting, put sufficient into a cold mould to cover the bottom of it. Then place in the centre, according to taste, any fine fruit you choose, a few grapes, cherries, strawberries, currants, anything you like, provided it is not too heavy to break the jelly. Put in another layer of jelly, and when it is set enough, a little more fruit, then fill up your mould with jelly, and let it stand for some hours.

RIBBON JELLY.

Soak one ounce of Nelson's Patent Gelatine in half-a-pint of cold water for twenty minutes, then add the same quantity of boiling water. Stir until dissolved, and add the juice and peel of two lemons, with wine and sugar sufficient to make the whole [67] quantity one quart. Have ready the white and shell of an egg, well beaten together, and stir these briskly into the jelly; then boil for two minutes without stirring, and remove it from the fire; allow it to stand two minutes, then strain it through a close flannel bag. Divide the jelly in two equal parts, leaving one pint of a yellow colour, and adding a few drops of prepared cochineal to colour the remainder a bright red. Put a small quantity of red jelly into a mould previously soaked in cold water. Let this set, then pour in a small quantity of the pale jelly, and repeat this until the mould is full, taking care that each layer is perfectly firm before pouring in the other. Put it in a cool place, and the next day turn it out. Or, the mould may be partly filled with the yellow jelly, and when this is thoroughly set, fill up with the red.

Ribbon jelly and jelly of two colours can be made in any pretty fancy mould (there are many to be had for the purpose); of course one colour must always be perfectly firm before the other is put in, or the effect would be spoilt by the two colours running into each other. Ribbon jelly can be made with two kinds of Nelson's Bottled Jelly. The Sherry will be used for the pale, and Cherry or Port Wine jelly for the red colour. Thus an elegant jelly will be made in a few minutes.

CLARET JELLY.

Take one ounce of Nelson's Patent Gelatine, soak for twenty minutes in half-a-pint of cold water, then [68] dissolve. Add three-quarters of a pound of sugar, a pot of red-currant jelly, and a bottle of good ordinary claret, and stir over the fire till the sugar is dissolved. Beat the whites and shells of three eggs, stir them briskly into the preparation, boil for two minutes longer, take it off the fire, and when it has stood for two minutes pass it through the bag. This should be a beautiful red jelly, and perfectly clear.

COFFEE JELLY.

Soak an ounce of Nelson's Gelatine in half-a-pint of water for an hour or more, dissolve it in a pint-and-a-half of boiling water with half-a-pound of sugar. Clear it with white of egg, and run through a jelly-bag as directed for making "a quart of brilliant jelly." This done, stir in a tablespoonful, or rather more if liked, of Allen and Hanbury's Café Vierge, which is a very fine essence of coffee. Or, instead of dissolving the Gelatine in water, use strong coffee.

COCOA JELLY.

Make half-a-pint of cocoa from the nibs, taking care to have it clear. Soak half-an-ounce of Nelson's Gelatine in half-a-pint of water; add a quarter of a pound of sugar, dissolve, and clear the jelly with the whites and shells of two eggs in the usual way. Flavour with Nelson's Essence of Vanilla after the jelly has been through the bag.

When a clear jelly is not required, the cocoa can be made of Schweitzer's Cocoatina, double the quantity required for a beverage being used. Mix this with half-an-ounce of Nelson's Gelatine and flavour with vanilla. [69]

ORANGES FILLED WITH JELLY.

Cut a small round from the stalk end of each orange, and scoop out the inside. Throw the skins into cold water for an hour to harden them, drain, and when quite dry inside, half fill with pink jelly. Put in a cool place, and when the jelly is firm, fill up with pale jelly or blanc-mange; set aside again, and cut into quarters before serving. Arrange with a sprig of myrtle between each quarter. Use lemons instead of oranges if preferred.

ORANGE FRUIT JELLY.

Boil half-a-pound of lump sugar in a gill of water until melted. Stir in half-an-ounce of Nelson's Gelatine previously soaked in a gill of cold water; when it is dissolved beat a little, and let it stand until cold. Rub four lumps of sugar on the peel of two fine oranges, so as to get the full and delicate flavour; add this sugar with the juice of a lemon and sufficient orange juice strained to make half-a-pint to the above. Beat well together, and when on the point of setting, add the fruit of two oranges prepared as follows: Peel the oranges, cut away all the white you can without drawing the juice, divide the orange in quarters, take out seeds and all pith, and cut the quarters into three or four pieces. Mix these with the jelly, which at once put into a mould, allowing it to stand a few hours before turning out.

APPLE JELLY.

Take one pound of apples, peel them with a sharp knife, cut them in two, take out the core, and cut the fruit into small pieces. Place the apples in a stewpan, [70] with three ounces of lump sugar, half-a-pint of water, a small teaspoonful of Nelson's Citric Acid, and six drops of Nelson's Essence of Lemon. Put the stewpan on the fire, and boil the apples till they are quite tender, stirring occasionally to prevent the fruit sticking to the bottom of the pan; or the apples can be steamed in a potato-steamer, afterwards adding lemon-juice and sugar. Soak an ounce of Nelson's Gelatine in a gill of cold water, dissolve it, and when the apples are cooked to a pulp, place a hair sieve over a basin and rub the apples through with a wooden spoon; stir the melted Gelatine into the apples, taking care that it is quite smoothly dissolved. If liked, colour part of the apples by stirring in half a spoonful of cochineal colouring.

Rinse a pint-and-a-half mould in boiling water, and then in cold water; ornament the bottom of the mould with pistachio nuts cut in small pieces, or preserved cherries, according to taste. When on the point of setting put the apples into the mould, and if any part of the apples are coloured, fill the mould alternately with layers of coloured and plain apples. Stand the mould aside in a cool place to set the apples, then turn out the jelly carefully on a dish, and send to table with cream whipped to a stiff froth.

LEMON SPONGE.

To an ounce of Nelson's Gelatine add one pint of cold water, let it stand for twenty minutes, then dissolve it over the fire, add the rind of two [71] lemons thinly pared, three-quarters of a pound of lump sugar, and the juice of three lemons; boil all together two minutes, strain it and let it remain till nearly cold, then add the whites of two eggs well beaten, and whisk ten minutes, when it will become the consistence of sponge. Put it lightly into a glass dish immediately, leaving it in appearance as rocky as possible.

This favourite sweetmeat is also most easily and successfully made with Nelson's Lemon Sponge. Dissolve the contents of a tin in half-a-pint of boiling water, let it stand until it is on the point of setting, then whip it until very white and thick.

If any difficulty is experienced in getting the Lemon Sponge out of the tin, set it in a saucepan of boiling water for fifteen minutes. In cold weather also, should the sponge be slow in dissolving, put it in a stewpan with the boiling water and stir until dissolved; but do not boil it. It is waste of time to begin whipping until the sponge is on the point of setting. A gill of sherry may be added if liked, when the whipping of the sponge is nearly completed. Put the sponge into a mould rinsed with cold water. It will be ready for use in two or three hours. A very pretty effect is produced by ornamenting this snow-white sponge with preserved barberries, or cherries, and a little angelica cut into pieces to represent leaves.

STRENGTHENING JELLY.

Put one ounce each of sago, ground rice, pearl barley, and Nelson's Gelatine—previously soaked in [72] cold water—into a saucepan, with two quarts of water; boil gently till the liquid is reduced one-half. Strain and set aside till wanted. A few spoonfuls of this jelly may be dissolved in broth, tea, or milk. It is nourishing and easily digested.

DUTCH FLUMMERY.

To an ounce and a half of Nelson's Patent Gelatine add a pint of cold water; let it steep, then pour it into a saucepan, with the rinds of three lemons or oranges; stir till the Gelatine is dissolved; beat the yolk of three eggs with a pint of good raisin or white wine, add the juice of the fruit, and three-quarters of a pound of lump sugar. Mix the whole well together, boil one minute, strain through muslin, stir occasionally till cold; then pour into moulds.

ASPIC JELLY.

Were it not for the trouble of making Aspic Jelly, it would be more generally used than it is, for it gives not only elegance but value to a number of cold dishes. We have now the means of making this with the greatest ease, rapidity, and cheapness. Soak an ounce of Nelson's Gelatine in a pint of cold water, dissolve it in a pint of boiling water, add a large teaspoonful of salt, a tablespoonful of French vinegar, and the contents of a tin of Nelson's Extract of Meat dissolved in a gill of boiling water. Wash the shell of an egg before breaking it, beat up white and shell to a strong froth, and stir into the aspic. Let it come [73] slowly to the boil, and when it has boiled two minutes, let it stand for another two minutes, then strain through a flannel bag kept for the purpose. If a stiff aspic is required, use rather less water.

HOW TO MAKE A JELLY-BAG.

The very stout flannel called double-mill, used for ironing blankets, is a good material for a jelly-bag. Take care that the seam of the bag be stitched twice, to secure the jelly against unequal filtration. The bag may, of course, be made any size, but one of twelve or fourteen inches deep, and seven or eight across the mouth, will be sufficient for ordinary use. The most convenient way of using the bag is to tie it upon a hoop the exact size of the outside of its mouth, and to do this tape should be sewn round it at equal distances.

If there is no jelly-bag in a house, a good substitute may be made thus: Take a clean cloth folded over corner-ways, and sew it up one side, making it in the shape of a jelly-bag. Place two chairs back to back, then take the sewn-up cloth and hang it between the two chairs by pinning it open to the top bar of each chair. Place a basin underneath the bag. Here is another substitute: Turn a kitchen stool upside down, and tie a fine diaper broth napkin, previously rinsed in hot water, to the four legs, place a basin underneath and strain through the napkin.

[74]

CREAMS.

The careful housekeeper of modern times has been accustomed to class creams among the luxuries which can only be given on special occasions, both because they take so much time and trouble to make, and because the materials are expensive. It is, nevertheless, possible to have excellent creams made on a simple plan and at a moderate cost. Cream of a superior kind is now everywhere to be had in jars, condensed milk answers well, and by the use of Nelson's Gelatine, and any flavouring or syrup, excellent creams can be made. Our readers will find that the method of the following recipes is simple, the cost moderate, and the result satisfactory. A hint which, if acted on, will save time and trouble, may be given to inexperienced persons intending to make creams similar to Lemon Cream, which is light and frothy. Do not add the lemon-juice until the mixture of cream and lemon-juice is nearly cold, and do not commence whipping until it is on the point of setting.

Delicious and inexpensive creams can be made by dissolving any of Nelson's Tablet Jellies in half the quantity of water given in the directions for making the jelly, and adding cream, either plain or whipped, in the same way as directed for Orange Cream and Cherry Cream.

[75]

LEMON CREAM.

Soak an ounce of Nelson's Gelatine in half-a-pint of milk, dissolve it in a pint of boiling milk with a quarter of a pound of lump sugar. When nearly cold, add a gill of lemon-juice and whisk the cream until it is light and sponge-like. Then stir in a gill of whipped cream, put into a mould, and let it stand for two or three hours.

Or, dissolve a pint tablet of Nelson's Lemon Tablet Jelly in half-a-pint of hot water. When cool, add to it half-a-pint of cream, and whisk together until on the point of setting, when mould it.

STRAWBERRY CREAM.

Dissolve an ounce of Nelson's Gelatine, previously soaked in a gill of cold water, in a pint of hot milk. When it is so nearly cold as to be on the point of setting, add half-a-pint of strawberry syrup, and sufficient rose colouring to make it a delicate pink; whisk the cream until it is light and frothy, stir in lightly a gill of whipped cream, then mould it.

A good syrup can be made for this cream by putting half-a-pound of strawberry and half-a-pound of raspberry jam into half-a-pint of boiling water, and, after having well stirred it, rubbing it through a fine sieve. The syrup should not be too sweet, and the addition of the juice of one or two lemons, or a little citric acid, will be an advantage.

Creams, which have cochineal colouring in them, should not be put into tin moulds, as this metal turns [76] them of a mauve shade. Breton's Rose Colouring is recommended, because it is prepared from vegetables, and is free from acid.

ORANGE CREAM.

Dissolve a pint tablet of Nelson's Orange Tablet Jelly in half-a-pint of hot water. When cool, mix with it half-a-pint of cream or milk, and whip together until the cream is on the point of setting.

IMITATION LEMON CREAM.

This will be found useful when cream is not to be had. Put the thin peel of two lemons into half-a-pint of boiling water, and when it has stood a little, dissolve half-a-pound of loaf sugar in it. When nearly cold, add three eggs, the yolks and whites well beaten together, and the juice of the lemons. Strain this into a stewpan, and stir until it is well thickened. After taking from the fire, stir occasionally until cold, then mix into it a quarter of an ounce of Nelson's Gelatine soaked and dissolved in half a gill of water, also nearly cold.

APRICOT CREAM.

Drain the juice from a tin of preserved apricots, add to it an equal quantity of water; make a syrup by boiling with this half-a-pound of lump sugar until it begins to thicken; then put in the apricots and simmer them gently for ten minutes. Drain away the syrup, and put both it and the fruit aside separately for use as directed.

[77] Dissolve an ounce of Nelson's Gelatine, previously soaked, in a quart of boiling milk lightly sweetened, and, when at the point of setting, put a teacupful of it gently into a mould, then a layer of the apricots; wait a minute or two before putting in another cup of cream, then fill up the mould with alternate layers of fruit and cream. Let the cream stand some hours before turning out, and when it is on its dish pour round it the syrup of apricots.

PINEAPPLE CREAM.

Drain the syrup from a tin of pineapple, boil it down to half. Cut the best part of the pineapple into neat little squares, pound the remainder, which press through a strainer. Make a custard with half-a-pint of milk and three yolks of eggs. Measure the quantity of syrup and fruit juice, and dissolve Nelson's Gelatine in the proportion of half-an-ounce to a pint of it and custard together. Mix the gelatine with the custard, then put in the pieces of pineapple, and when it is cold the syrup, the juice, and two tablespoonfuls of whipped cream. Have ready a little of Nelson's Bottled Cherry or Port Wine Jelly melted in a fancy mould, which turn round so that it adheres to the sides, and when the first quantity is set, put in a little more. As the cream is on the point of setting, put it into the mould and allow it to stand until firm. When turned out, ornament the cream with the remainder of the bottled jelly lightly chopped.

PALACE CREAM.

Make a custard of three eggs and a pint-and-a-half [78] of milk sweetened, when it is ready dissolve in it an ounce of Nelson's Gelatine, previously soaked in half-a-pint of milk. When made, the quantity of custard should be fully a pint-and-a-half, otherwise the cream may be too stiff. When the cream is cool, put a little into a mould, previously ornamented with glacé cherries and little pieces of angelica to represent leaves. The fruit is all the better if soaked in a little brandy, as are the cakes, but milk can be used for these last. Put a portion of two ounces of sponge-cakes and one ounce of ratafias on the first layer of cream, keeping it well in the centre, and then fill up the mould with alternate layers of cakes and cream. When turned out, a little liqueur or any kind of syrup can be poured round the cream.

FRUIT CREAM.

Strain the juice from a bottle of raspberries and currants on to three-quarters of a pound of loaf sugar, boil up, then simmer for half-an-hour. Mix the fruit and a large tablespoonful of raspberry jam with the syrup, and rub it through a hair sieve. Dissolve Nelson's Gelatine, in the proportion of half-an-ounce to a pint of the fruit, in a little water, stir well together. When cold put it into a border mould, and as soon as it is firm turn out and fill the centre with a cream, which make with half-an-ounce of Nelson's Gelatine and three gills of milk, sweetened and flavoured with Nelson's Essence of Vanilla. Whisk until cool, when stir in a gill of whipped cream.

MANDARIN CREAM.

Dissolve half-an-ounce of Nelson's Gelatine, pre [79] viously soaked in half-a-pint of cold milk, in half-a-pint of sweetened boiling milk or cream. Dissolve a pint bottle of Cherry Jelly as directed. When the last is on the point of setting put a layer into a mould, then a layer of the cream, each of these about an inch deep, and fill up the mould in this way. This quantity of material will make two handsome moulds, suitable for a supper party.

BLANC-MANGE.

To an ounce of Nelson's Gelatine add half-a-pint of new milk, let it soak for twenty minutes, boil two or three laurel leaves in a pint of cream and half-a-pint of milk; when boiling pour over the soaked gelatine, stir it till it dissolves, add four or five ounces of lump sugar and a little brandy if approved; strain it through muslin, stir occasionally till it thickens, and then put it into moulds.

SOLID SYLLABUB.

Soak an ounce of Nelson's Gelatine twenty minutes in three-quarters of a pint of water, add the juice and peel of two large lemons, a quarter of a pint of sherry, five or six ounces of lump sugar; boil the above two minutes, then pour upon it a pint of warm cream, stir it quickly till it boils, then strain and stir till it thickens, and pour it into moulds.

CHARLOTTE RUSSE.

Line a plain mould at the bottom and sides with sponge finger-biscuits, fill it with strawberry cream, or cream made as directed in the several recipes. If the weather is warm it will be necessary to place the Charlotte on ice for an hour or two, but in the winter it will turn out without this. The biscuits for a [80] Charlotte Russe should be made quite straight, and in arranging them in the mould they should lap slightly one over the other.

BADEN-BADEN PUDDING.

Dissolve an ounce of Nelson's Gelatine, previously soaked in half-a-pint of cold milk, in a pint-and-a-half of boiling milk; when it is nearly cold stir into it an ounce of rice, well boiled or baked; flavour the pudding to taste, and when on the point of setting put it into a mould and let it stand for two or three hours; serve plain or with stewed fruit.

CHERRY CREAM.

Dissolve a pint tablet of Nelson's Cherry Tablet Jelly in half-a-pint of hot water. When cool, mix with it half-a-pint of cream or milk, and whip together until the cream is on the point of setting.

VELVET CREAM.

Soak three-quarters of an ounce of Nelson's Patent Gelatine in half-a-pint of sherry or raisin wine, then dissolve it over the fire, stirring all the time; rub the rinds of two lemons with six ounces of lump sugar, add this, with the juice, to the hot solution, which is then to be poured gently into a pint of cream; stir the whole until quite cold, and then put into moulds.

This can be made with a pint of boiling milk, in which an ounce of Nelson's Gelatine, previously soaked in half-a-pint of cold milk, has been dissolved, and flavoured and sweetened.

[81]

ITALIAN CREAM.

Take three-quarters of an ounce of Nelson's Patent Gelatine and steep it in half-a-pint of cold water; boil the rind of a lemon, pared thinly, in a pint of cream; add the juice of the lemon and three tablespoonfuls of raspberry or strawberry syrup to the soaked Gelatine; then pour the hot cream upon the above ingredients, gently stirring the while. Sweeten to taste, and add a drop or two of prepared cochineal. Whisk till the mixture is thick, then pour into moulds.

CHEESE AND MACARONI CREAM.

Boil two ounces of macaroni, in water slightly salted, until tender, when drain; cut it into tiny rings, and put it into a stewpan with half-a-pint of milk or cream, keeping it hot on the stove without boiling for half-an-hour. Soak and dissolve half-an-ounce of Nelson's Gelatine in half-a-pint of milk, and when this and the macaroni are cold, stir together, add two ounces of grated Parmesan cheese, with salt and cayenne pepper to taste. Stir occasionally until the cream is on the point of setting, when mould it. Should the cream be absorbed by the macaroni, more must be added to bring the whole quantity of liquid to one pint. If preferred, rice well boiled or baked in milk, or vermicelli paste, can be substituted for the macaroni.

COFFEE CREAM.

Dissolve an ounce of Nelson's Gelatine, previously [82] soaked in half-a-pint of cold milk, in a pint-and-a-half of boiling milk with two ounces of sugar; stir in sufficient strong Essence of Coffee to flavour it, and when on the point of setting put it into a mould.

CHOCOLATE CREAM.

Boil a quarter of a pound of loaf sugar in a pint of milk. Dissolve in it an ounce of Nelson's Gelatine, previously soaked in half-a-pint of cold milk, and stir into it three teaspoonfuls of Schweitzer's Cocoatina, dissolved in half-a-pint of boiling milk. Beat until on the point of setting, and put the cream into a mould. A few drops of Nelson's Essence of Vanilla can be added with advantage.

CHARTREUSE OF ORANGES.

Peel four or five oranges, carefully take out the divisions which put on a hair sieve in a cool place to drain all night. Melt a little Nelson's Bottled Orange Jelly, pour it into a saucer and dip in each piece of orange, which arrange in a close circle round the bottom of a small pudding-basin. Keep the thick part of the orange downwards in the first row, in the next put them the reverse way. Continue thus until the basin is covered. Pour in a little of the melted jelly, then of cream, made by mixing a quarter of an ounce of Nelson's Gelatine soaked and dissolved in a gill of milk, into a gill of rich cream, sweetened. Fill up the basin with alternate layers of jelly and cream, allowing each of these to set before the other is put in, making the jelly layers last. The [83] Chartreuse will turn out easily if the jelly is gently pressed from the basin all round. Garnish with two colours of Nelson's Bottled Jelly lightly chopped.

FIG CREAM.

Preserved green figs are used for this cream — those of Fernando Rodrigues are excellent. Place the figs in a plain mould, and pour in gently, when on the point of setting, a cream made with a pint of cream and half-an-ounce of Nelson's Gelatine, and lightly sweetened. When the cream is turned out of the mould, pour round it the syrup in which the figs were preserved.

CHAMPAGNE CREAM.

Although this is properly a jelly, when well made it eats so rich that it is usually called cream. It is chiefly used in cases of illness, when it is desirable to administer champagne in the form of jelly. Soak half-an-ounce of Nelson's Gelatine in a gill of cold water, dissolve it in a stewpan with one or two ounces of sugar, according as the jelly is required sweet or otherwise. When cool, add three gills of champagne and two tablespoonfuls of lemon juice, whip until it is beginning to set and is light and frothy; put into a mould, and it will be ready for use in two hours, if put in a cold place.

ORANGE MOUSSE.

Rub the zest of the peel of two oranges on to a quarter of a pound of lump sugar, which boil with half-a-gill of water to a thick syrup. Beat the juice of three large oranges with two whole eggs, and having [84] whisked them slightly, add the syrup and Nelson's Gelatine, dissolved, in the proportion of half-an-ounce to a pint of liquid. Whisk the mixture over a saucepan of hot water until it is warm, then place the basin in another with cold water and continue whisking until it is beginning to set, when put it into a fancy mould.

STRAWBERRY TRIFLE.

Put a layer of strawberry jam at the bottom of a trifle dish. Dissolve a half-pint tablet of Nelson's Raspberry Jelly, and when it is set break it up and strew it over the jam. Upon this lay sponge finger biscuits and ratafia cakes, and pour over just enough new milk to make them soft. Make a thick custard, flavoured with Nelson's Essence of Vanilla, and spread it over the cakes. Finally place on the top a handsome quantity of cream, whisked with a little powdered sugar and flavoured with vanilla.

WHIPPED CREAM.

To half-a-pint of cream put a tablespoonful of fine sifted sugar, add sufficient of any of Nelson's Essences to give it a delicate flavour. With a whisk or wire spoon, raise a froth on the cream, remove this as soon as it rises, put it on a fine hair, or, still better, lawn sieve; repeat this process until the cream is used up. Should the cream get thick in the whisking, add a very little cold water. Put the sieve containing the whisked cream in a basin and let it stand for some hours, which will allow it to become more solid and fit for such purposes as filling meringues.

[85]

CAKES.

The proper beating of the whites of the eggs is an important matter in cake-making. There are a number of machines for this purpose, which are in turn eagerly adopted by inexperienced persons; but for private use not one of them is comparable to hand-beating. When once the knack of beating eggs is acquired but little labour is needed to bring them to the right consistency; indeed, the most successful result is that which is the most rapidly attained. The whites of eggs for beating should be fresh, and should be carefully separated from the yolks by passing and repassing them in the two halves of the shell. It is best to beat the whites immediately they are broken, but if this is not possible, they must be kept in a cool place until wanted. If ice is at hand, it will be found advantageous to keep the eggs in it. In well-furnished kitchens a copper beating-bowl is provided; it should not be tinned, as contact with this metal will blacken the eggs; for this reason, the whisk, if of iron wire, should not be new. An earthenware bowl with circular bottom, and sufficiently large to admit of a good stroke in beating, answers the purpose perfectly well. A pinch of salt may be added to the whites, and if an inexperienced beater finds them assume a granulated appearance, a little lemon-juice will remedy it.

[86] Begin by beating gently, increasing the pace as the egg thickens. As it is the air mixing with the albumen of the eggs which causes them to froth, it is necessary to beat them in a well-ventilated and cool place, so that they may absorb as much air as possible.

If these simple and important conditions are observed, the whites of a dozen eggs may be beaten to the strongest point, without fatigue to the operator, in five minutes. When the whites are properly beaten they should turn out of the bowl in one mass, and, after standing a little while, will not show signs of returning to their original state.

In order more easily to make cakes and biscuits into the composition of which almonds and cocoa-nut enter largely, manufacturers supply both of these pounded or desiccated. It is, however, preferable to prepare the former fresh, and much time and trouble may be saved in passing almonds through Kent's Combination Mincer, 199, High Holborn, instead of laboriously pounding them in a mortar. The result is, besides, more satisfactory, the paste being smoother than it can otherwise be made in domestic practice.

Cakes of the description for which we now give recipes cannot be made well unless the materials are properly prepared and thoroughly beaten. It is clear that if eggs are not beaten to such a consistency that they will bear the weight of the other ingredients, the result must be a heavy cake.

Currants for cakes, after they have been washed and picked, should be scalded, in order to swell them and make them more tender.

[87] Put the currants into a basin, pour boiling water over them, cover the basin with a plate; after they have stood a minute, drain away the water and throw the fruit on a cloth to absorb the moisture. Put the currants on a dish or plate in a very cool oven, turning occasionally until thoroughly dry; dust a little flour over them, and they will be ready for use.

Castor sugar for cakes works more easily when it is fine. For superior cakes raw sugar will not answer.

POUND CAKE.

One pound fresh butter, one pound Vienna flour, six eggs (or seven, if small), one pound castor sugar, quarter of a pound almonds cut small, half-a-pound of currants or sultanas, three ounces of candied peel, a few drops of essence of ratafia.

The butter to be beaten to a cream. If it is hard warm the pan. Add the sugar gradually; next the eggs, which must previously be well beaten up; then sift in the flour; and, last of all, put in fruit, almonds, and flavouring.

This cake takes about half-an-hour to mix, as all the ingredients must be well beaten together with an iron spoon from left to right. Bake in small tins, for about forty minutes, in a moderate oven.

PLAIN POUND CAKE.

Half-a-pound of fresh butter, three eggs, one pound of Vienna flour, one pound of castor sugar, a quarter of a pound of almonds cut small, half-a-pound of [88] currants, three ounces of candied peel, a few drops of essence of ratafia.

Beat the butter to a cream, from left to right, and mix in the sugar gradually. Beat the eggs up, and mix them with half-a-pint of new milk; stir into the butter; then add the flour; and, last of all, the fruit.

SAVOY SPONGE CAKE.

Beat half-a-pound of finely sifted sugar with the yolks of four eggs until you have a thick batter, stir in lightly six ounces of fine dry sifted flour, then the whites of the eggs beaten to a very strong froth. Have ready a tin which has been lightly buttered, and then covered with as much sifted sugar as will adhere to it. Pour in the cake mixture, taking care the tin is not more than half full, and bake for half-an-hour.

LEMON SAVOY SPONGE.

Half-a-pound of loaf sugar, rub some of the lumps on the peel of two lemons, so as to get all the flavour from them; dissolve the sugar in half a gill of boiling water; add the juice of the lemons, or one of them if a large size, and beat with the yolks of four eggs until very white and thick; stir in a quarter of a pound of fine flour, beat the whites of the eggs to a strong froth, and mix as thoroughly but as lightly as possible; butter and sift sugar over a mould, nearly fill it with cake mixture, and bake at dark yellow paper heat for thirty minutes.

[89]

MACAROONS.

Beat up a packet of Nelson's Albumen with three teaspoonfuls of cold water to a strong froth, mix in half-a-pound of finely-sifted sugar and two ounces each of pounded sweet and bitter almonds. Flour a baking-sheet, and lay on it sheets of wafer-paper, which can be bought at the confectioner's, and drop on to them at equal distances, a small piece of the paste. Bake in a moderate oven for ten minutes, or until the macaroons are crisp and of a golden colour. When done cut round the wafer-paper with a knife, and put the cakes on a sieve to dry.

In following recipes for this class of cake some judgment is required in the choice of the sugar, and the result will vary greatly according as this is of the right sort, or otherwise. A little more or less sugar may be required, and only practice can make perfect in this matter. As a general direction, it may be given that the sugar must be of the finest quality, and be very finely sifted, but not flour-like.

COCOA-NUT CAKES.

Beat up a packet of Nelson's Albumen with three teaspoonfuls of cold water to a strong froth, mix with it a quarter of a pound of finely sifted sugar, and two ounces of Edwards' Desiccated Coker-nut. Put sheets of wafer-paper on a baking-tin, drop small pieces of the cake mixture on to it, keeping them in a rocky shape. [90] Bake in a moderate oven for ten minutes, or until crisp.

CHOCOLATE CAKES.

Whisk a packet of Nelson's Albumen with three teaspoonfuls of cold water to the strongest possible froth, mix in half-a-pound of finely sifted sugar, two teaspoonfuls of Schweitzer's Cocoatina, and six drops of Nelson's Essence of Vanilla; sift paper thickly with sugar, and drop small teaspoonfuls of the mixture at equal distances on it, allowing space for the cakes to spread a little. Bake for ten minutes in a moderate oven.

COCOA-NUT ROCK.

Boil half-a-pound of loaf sugar in a gill of water until it is beginning to return again to sugar, when cool add a packet of Nelson's Albumen whisked to a strong froth with three teaspoonfuls of water, and stir in a quarter of a pound of Edwards' Desiccated Coker-nut. Spread the mixture, not more than an inch thick, in a greased pudding-tin, and place in a cool oven to dry. When done cut in neat squares, and keep in tins in a cool, dry place.

SUGAR ICING.

No icing can be successfully done unless the sugar is of the finest kind, perfectly white, and so finely sifted as hardly to be distinguished by the eye from potato-flour. Such sugar can now generally be pro [91] cured of the best grocers at a moderate price. The process of sifting the sugar at home is somewhat slow and troublesome, but by so doing a perfectly pure article is secured. After being crushed the sugar should be passed through sieves of varying fineness, and, finally, through one made for the purpose, or failing this, very fine muslin will answer. When the sugar has been sifted at home, and it is certain there is no admixture of any kind with it, a small quantity of "fécule de pommes de terre" (potato-flour) may be added; it reduces sweetness, and does not interfere with the result of the process. If the sugar is not sifted very fine a much longer time will be required to make the icing, and in the end it will not look so smooth as it ought to do. Confectioners use pyroligneous acid instead of lemon-juice, and there is no objection to it in small quantities. To make the icing, beat up a packet of Nelson's Albumen dissolved with three teaspoonfuls of cold water, work in by degrees one pound of fine icing sugar, adding a teaspoonful of lemon-juice or a few drops of pyroligneous acid, which will assist in keeping the icing white, or a slight tinge of stone-blue will have the same effect. If potato-flour is used, mix it thoroughly with the sugar before adding it to the white of egg. A little more or less sugar may be required, as the result is in great measure determined by the method of the operator; and when the paste is perfectly smooth, and will spread without running, it is fit for use. For icing large cakes confectioners use a stand which has a revolving board, so that cakes can conveniently be turned about; failing [92] this, an ordinary board or inverted plate can be made to answer. As soon as the icing is spread on the cake it must be dried in an oven with the door open. It is sometimes found sufficient to keep the cake in a hot room for some hours. If too great heat is used the icing will crack.

ALMOND PASTE.

Blanch one pound of sweet and two ounces of bitter almonds, pound them in a mortar, adding a little rose-water as you go on, to prevent oiling; and when all the almonds are reduced to a perfectly smooth paste, mix them with an equal weight of icing sugar. Moisten the paste with a packet of Nelson's Albumen dissolved in three teaspoonfuls of cold water, and spread it evenly on the cake, allowing it to become dry and firm before spreading the icing over it. This paste can be used for making several kinds of cakes and sweetmeats, and without the Albumen can be kept in bottles for some time. Almond paste can be made from bitter almonds which have been infused in spirit to make an extract for flavouring, and in this case no sweet almonds will be required.

[93]

BEVERAGES.

Among the most useful preparations which have ever been introduced to the public for the immediate production of delicious beverages, are Nelson's Bottled Jellies. These beverages are highly approved for ordinary use at luncheon and dinner, as well as for afternoon and evening entertainments, and have a special value for invalids, as they contain nourishment and are at the same time very refreshing. When required for use, dissolve a bottle of the jelly, and mix with it five times its bulk of water, the beverage can then be used either hot or cold; if in standing it should be slightly thickened it will only be necessary briskly to stir it with a spoon. Lemon, orange, and cherry jelly, with the addition of water as directed, will be found superior to any other beverage of the kind, and specially excellent for children's parties.

The following "cups" are delicious made with the jelly as directed.

Claret Cup, made merely with seltzer water, claret, and Port Wine Jelly, will be found superior to the ordinary preparation. A little sugar may be added if desired. To a bottle of claret and a pint of seltzer-water use a half-pint bottle of Port Wine Jelly, stir briskly until well mixed, put in a sprig of balm and borage, three thick slices of cucumber; place the vessel [94] containing the claret cup covered over on ice for an hour; strain out the herbs before serving.

Badminton Cup is made with Burgundy, in the same way as the above, with the addition of a bottle of Orange Jelly.

Champagne Cup requires equal quantities of the wine and seltzer-water, with a bottle of Orange Jelly.

Cider Cup is made with a pint and a half of cider, a bottle of soda-water, and a bottle of either Orange, Lemon, or Sherry Jelly.

Cherry Cup.—Half-a-pint of claret, a quart of soda-water, and a half-pint bottle of Cherry Jelly.

MULLED PORT WINE.

Dissolve a bottle of Port Wine Jelly and add to it four times its bulk of boiling water with a little nutmeg, and, if liked, a crushed clove.

LEMONADE.

Half-a-teaspoonful of Nelson's Citric Acid dissolved in a quart of water, with a sliced lemon and sweetened with sugar, forms a good lemonade, and is a cooling and refreshing drink. A small pinch of the Citric Acid dissolved in a tumbler of water with a little sugar and a pinch of bicarbonate of potash, makes an effervescing draught. These acidulated drinks are exceedingly useful for allaying thirst; and as refrigerants in feverish and inflammatory complaints they are invaluable.

[95]

LEMONADE (A NEW RECIPE).

Dissolve three-quarters of a pound of loaf sugar and the contents of a threepenny packet of Nelson's Citric Acid in a quart of boiling water; then add two quarts of fresh cold water and one-twelfth part of a bottle of Nelson's Essence of Lemon. The above quantity of sugar may be increased or decreased according to taste.

GINGERADE.

Crush an ounce of whole ginger, pour over it a quart of boiling water, cover the vessel, and let the infusion stand until cold. (The Extract of Ginger may be used in place of this infusion). Strain through flannel; add a teaspoonful of Nelson's Citric Acid, six drops of Nelson's Lemon Flavouring, and a quarter of a pound of lump sugar; stir until dissolved, and the Gingerade will be ready.

AN EXTRACT OF GINGER FOR FAMILY USE.

An Extract of Ginger made as follows is most useful for family purposes, and can be substituted for the infusion in Gingerade. Crush half-a-pound of fine whole ginger in the mortar, or cut into small pieces. Put into a bottle with half-a-pint of unsweetened gin, let it stand for a month, shaking it occasionally, then drain it off into another bottle, allowing it to stand until it has become clear, when it will be fit for use.

LEMON SYRUP.

Boil a pound of fine loaf sugar in a pint-and-a-half of water. Remove all scum as it rises, and continue boiling gently until the syrup begins to thicken and assumes a golden tinge, then add a pint of strained lemon-juice or a packet of Nelson's Citric Acid dissolved in water, and allow both to boil together for half-an-hour. Pour the syrup into a jug, to each pint add one-twelfth part of a bottle of Nelson's Essence of Lemon, and when cold bottle and cork well.

The juice of Seville oranges may be made into a syrup in the same way as that of lemons, or lemon and orange juice may be used in equal quantities. These syrups are useful for making summer drinks, and for invalids as lemonade or orangeade.

MILK BEVERAGE.

A very agreeable and useful beverage is made by dissolving a quarter of an ounce of Nelson's Gelatine in a pint of milk. A spoonful of cream can, if preferred, be used with a bottle of soda-water. For invalids, this beverage can be used instead of tea or coffee, and may be preferable in many cases on account of the nourishment it contains; it will also be found an excellent substitute, taken hot, for wine-whey, or posset, as a remedy for a cold. For summer use, Milk Beverage is delicious, and may be flavoured with raspberry or strawberry syrup. If on standing it should thicken, it will only be necessary briskly to beat it up with a spoon.

CITRIC ACID.

This acid exists in the juice of many fruits, such as the orange, currant, and quince, but especially in that of the lemon. It is chiefly made from the concentrated juice of lemons, imported from Sicily and Southern Italy, and which, after undergoing certain methods of preparation, yields the crystals termed Citric Acid. These crystals may be used for all the purposes for which lemon-juice is employed. In the manufacture of the Citric Acid now offered to the public by Messrs. G. Nelson, Dale, and Co., only the pure juice of the lemon is used.

ESSENCE OF LEMON.

This well-known essence is extracted from the little cells visible in the rind of lemons, by submitting raspings of the fruit to pressure. The greater portion of the oil of lemons sold in England is imported from Portugal, Italy, and France. It is very frequently adulterated with oil of turpentine. In order to present the public with a perfectly pure commodity, G. Nelson, Dale, and Co. import their Essence of Lemon direct from Sicily, and from a manufacturer in whom they have the fullest confidence.

Nelson's Essence of Lemon is sold in graduated bottles, eightpence each, each bottle containing sufficient for twelve quarts of jelly.

[98]

MACARONI, ETC.

We now give recipes for a few useful little dishes, chiefly of macaroni, which can be had at such a price as to bring it within the reach of all classes. English-made macaroni can be bought at fourpence, and even less, the pound, and the finest Italian at sixpence. The Naples, or pipe-macaroni, is the most useful for families, and the Genoa, or twisted, for high-class dishes. The English taste is in favour of macaroni boiled soft, and in order to make it so, many cooks soak it. But this is not correct, and it is not at all necessary to soak macaroni. If kept boiling in sufficient water, the macaroni requires no attention—ebullition prevents it sticking to the saucepan.

Although we give several ways of finishing macaroni, it is excellent when merely boiled in water with salt, as in the first recipe, eaten as an accompaniment to meat, or with stewed fruit.

MACARONI WITH CHEESE.

Throw a quarter of a pound of macaroni broken into pieces an inch long, into three pints of boiling water, with a large pinch of salt. The saucepan should be large, or the water will rise over when the macaroni boils fast, which it should do for twenty or twenty-five minutes. When done, strain the macaroni through a colander, put it back into the saucepan [99] with an ounce of fresh butter, a small pinch of white pepper and of salt, if necessary, and shake it over the fire for a minute or two. Take the saucepan off the fire, and stir into the macaroni two ounces or more, if liked, of grated Parmesan cheese. Serve immediately with crisp dry toast, cut in neat pieces. If not convenient to use Parmesan, a mild dry English or American cheese will answer very well. Some cooks prefer, when the macaroni is boiled, to put a fourth part of it on to a hot dish, then to strew over it a fourth part of the grated cheese, and so on till all of both are used, cheese, of course, covering the top.

MACARONI CHEESE.

Boil and drain the macaroni, mix with a quarter of a pound an ounce of butter, and two ounces of grated cheese; pepper or cayenne pepper and salt to taste. Put the macaroni in a dish and strew over it sufficient grated cheese to cover it up, run a little dissolved butter over the top, and put it in the oven till it is a bright-yellow colour; serve quickly.

MACARONI WITH BACON.

Boil two ounces of streaky bacon, cut it into dice or mince it, stir it into a quarter of a pound of macaroni boiled as for macaroni cheese: if liked, add a few drops of vinegar, pepper, and salt, and serve very hot.

MACARONI WITH ONIONS.

Boil the macaroni as above, mix with it two or three onions sliced and fried a delicate brown, add a few spoonfuls of gravy, stir over the fire for a few minutes and serve.

STEWED MACARONI.

Throw a quarter of a pound of macaroni into three pints of boiling water with a teaspoonful of salt, and let it boil for twenty minutes. Drain in a colander, then put it into a stewpan with half a tin of Nelson's Extract of Meat dissolved in half-a-pint of water, and stir over the fire for five minutes. Take it off the fire and stir in one ounce of grated cheese, pepper and salt to taste.

MACARONI WITH TOMATOES.

Prepare the macaroni as in the above recipe, put it into a stewpan with a small piece of butter and a teacupful of tomato sauce, or a small bottle of conserve of tomatoes, and stir briskly over the fire for five minutes.

SWEET MACARONI.

Boil the macaroni as for the other dishes, but with only a pinch of salt, until tender, when drained put into a stewpan with a gill of milk to each two ounces, and two ounces of sifted lump sugar. Any flavouring may be used, but perhaps there is nothing better than grated lemon-peel, and for those who like it, powdered cinnamon or grated nutmeg. Stir over the [101] fire until all the milk is absorbed; a little cream is, of course, an improvement. For those who do not like milk, the juice of a lemon, or a little sherry, may be substituted, and for a superior dish vanilla can be used for flavouring.

BOILED CHEESE.

Put four tablespoonfuls of beer into a small saucepan, shred into it a quarter of a pound of good new cheese, and stir briskly over the fire until all is dissolved and is on the point of boiling, then take it off instantly, for, if the cheese is allowed to boil, it will become tough. Have ready slices of toasted bread, spread the cheese on it, and serve as quickly as possible.

LES CANAPÉS AU PARMESAN.

Take the crumb of a French roll, cut it into rounds a quarter of an inch thick, put them into a wire frying-basket, immerse in hot fat, and crisp the bread instantly. Throw it on to paper, dry, and sprinkle over each piece a thick layer of grated Parmesan cheese, pepper, and salt. Put the canapés in a Dutch oven before a clear fire, just to melt the cheese, and serve immediately they are done.

RICE WITH PARMESAN CHEESE.

Boil a quarter of a pound of Patna rice in water with salt; drain it, toss it up in a stewpan with two ounces of fresh butter, and a pinch of cayenne [102] pepper. Put a quarter of the rice on a hot dish, strew over it equally an ounce of grated Parmesan cheese, then put another portion of rice and cheese until all is used. Serve immediately.

SCALLOPED EGGS.

Take a cupful of finely-sifted bread-crumbs, moisten them with a little cold milk, cream, or gravy, and season nicely with pepper and salt. Put a thin layer of the moistened crumbs on a lightly-buttered dish, cut two hard eggs into slices, and dip each piece in very thick well-seasoned white sauce, or Nelson's Extract of Meat dissolved in a little water, so as to glaze the eggs. Having arranged the slices of egg neatly on the layer of moistened bread-crumbs, cover them with another layer of it, and on the top strew thickly some pale gold-coloured raspings. Bake in a moderate oven for ten minutes. If potatoes are liked, they make a nice substitute for bread-crumbs. Take some mashed potatoes, add to them a spoonful of cream or gravy, and proceed as with bread-crumbs. Serve gravy made of Nelson's Extract of Meat with this dish.

SCOTCH WOODCOCK.

Melt a small piece of butter the size of a nut in a stewpan, break into it two eggs, with a spoonful of milk or gravy, and pepper and salt, stir round quickly until the eggs begin to thicken, keep the yolks whole as long as you can. When finished, pour on to a [103] buttered toast, to which has been added a little essence of anchovy or anchovy paste, and serve.

MUSHROOMS WITH CREAM SAUCE.

Dissolve two ounces of butter in a stewpan, mix in the yolks of two eggs lightly beaten, the juice of a lemon, and a pinch of pepper and salt, stir this over the fire until thickened. Have ready half-a-pint of plain butter sauce, and mix all gradually together, with a small tin of champignons, or about the same quantity of fresh mushrooms chopped and stewed gently for ten minutes in a little broth or milk. Stir them with the liquor in which they have stewed into the sauce, and let them stand for a few minutes, then spread the mixture on to neat slices of toasted bread. The sauce must be a good thickness, so that it will not run off the toast, and care must be taken in the first process not to oil the butter or make the sauce lumpy.

TO BOIL RICE (A BLACK MAN'S RECIPE).

As rice is so often badly cooked, we make no apology for giving the black man's celebrated recipe. Although he does not recommend a little salt in the water, we think that a small quantity should always be used, even when the rice has to be served as a sweet dish. "Wash him well, much wash in cold water, rice flour, make him stick. Water boil all ready, very fast. Shove him in; rice can't burn, water shake him too much. Boil quarter of an hour [104] or little more. Rub one rice in thumb and finger; if all rub away him quite done. Put rice in colander, hot water run away. Pour cup of cold water on him, put back in saucepan, keep him covered near the fire, then rice all ready. Eat him up."

TO MINCE VEGETABLES.

Peel the onion or turnip, put it on the board, cut it first one way in slices, not quite through, lest it should fall to pieces, then cut it in slices the other way, which will produce long cubes. Finally turn the onion on its side and cut through, when it will fall into dice-like pieces. The inconvenience and sometimes positive pain caused to the eyes by mincing or chopping the onions on a board is thus obviated, and a large quantity can be quickly prepared in the above way.

[105]

HINTS ON HOUSEKEEPING.

How many people are crying, "How can we save? Where can we retrench? Shall the lot fall on the house-furnishing, or the garden, or the toilet, or the breakfast or the dinner table? Shall we do with one servant less, move into a cheaper neighbourhood, or into a smaller house? No, we cannot make any such great changes in our way of life. There are the boys and girls growing up; we must keep up appearances for their sakes. We remember the old proverb that, 'however bad it may be to be poor, it is much worse to look poor.'" Yet, although, for many reasons, it is often most difficult to retrench on a large scale, there are people who find it easier, for instance, to put down the carriage than to see that the small outgoings of housekeeping are more duly regulated. It is seldom, indeed, that a wife can assist her husband save by lightening his expenses by her prudence and economy. Too many husbands, nowadays, can vouch for the truth of the old saying, "A woman can throw out with a spoon faster than a man can throw in with a shovel." The prosperity of a middle-class home depends very much on what is saved, and the reason that this branch of a woman's business is so neglected is that it is very difficult and very troublesome.

[106] "Take care of your pence and the pounds will take care of themselves," is a maxim that was much in use when we were young. Nowadays it is more fashionable to speak of this kind of thing as "penny wise and pound foolish." Looking to the outgoings of pence is voted slow work, and it is thought fine to show a languid indifference to small savings. "Such a fuss over a pennyworth of this or that, it's not worth while." Yes, but it is not that particular pennyworth which is alone in question, there is the principle involved—the great principle of thrift—which must underlie all good government. The heads of households little think of what evils they perpetuate when they shut their eyes to wasteful practices, because it is easier to bear the cost than to prevent waste.

The young servant trained under one careless how she uses, or rather misuses, that which is entrusted to her, carries in her turn the wasteful habits she has learned into another household, and trains others in a contempt for thrifty ways, until the knowledge of how to do things at once well and economically is entirely lost.

We often hear it urged that it is bad for the mind of a lady to be harassed by the petty details of small savings, and that if she can afford to let things go easily she should not be so harassed. But under no circumstances must any mistress of a household permit habitual waste in such matters. When the establishment is so large as to be to a great extent removed from the immediate supervision of the mistress, all she can do is to keep a careful watch [107] over every item of expenditure, and by every means in her power to let her servants feel that it is to their interest as well as to her own to keep within due bounds. A good cook is always a good manager. She makes many a meal of what an inferior cook would waste. The housekeeper should therefore insist on having good cooking at a reasonable cost, and never keep a cook who does not make the most of everything. In a large household a mistress cannot look after the sifting of cinders, but she can check her coal bills, and by observation find out in what department the waste is going on. It may not be possible to pay periodical visits to the gas-meter to see if the tap is turned on to the full when such force is not necessary, but she can from quarter to quarter compare notes, or have fixed, where it is easy for her to get at it, one of the gas-regulators now in use. And thus, by the exercise of judicious control and supervision, the guiding mind of the mistress will make itself felt in every department of the household without any undue worry to herself. The mistress of a small household who has things more under her immediate control, and whose income, no less than her sense of moral obligation, obliges her to look carefully after the outgoings, need not be told what a trial it is to be constantly on the watch to prevent waste. Probably she is compelled to leave a certain quantity of stores for general use; indeed, we doubt very much if there is anything saved by the daily giving out of ounces and spoonfuls of groceries, for if a servant is disposed to [108] be wasteful, she will be equally so with the small as the larger quantity.

What perpetual worry is caused by seeing how soap is left in the water until it is so soft as to have lost half its value! How many pence go in most households in that way every week, we wonder!

The scrubbing-brush also is left in water with the soap. A fairly good brush costs at least two shillings, and as one so treated only lasts half the proper time you may safely calculate that a shilling is soon wasted in that way. Brushes of all sorts are, as a rule, most carelessly used, and left about anyhow instead of being hung up. How much loss there is in a year in the careless use of knives and plate! Whenever possible both of these get into the hands of the cook. Her own tools from neglect or misuse have become blunt or worse, and she takes the best blade and the plated or silver spoon whenever she has a chance.

The plate gets thrown in a heap into an earthenware bowl to be bruised and scratched. The knives are either put insufficiently wiped through the cleaner, which is thus spoiled and made fit rather to dirty than clean knives, or they are left lying in hot water to have the handles loosened and discoloured.

Probably jars, tin boxes, and canisters are provided in sufficient quantity to put away and keep stores properly. But for all that, as it would seem in a most ingenious manner, loss and waste are contrived. Raw sugar is kept in the paper until it rots through it. Macaroni, rice, and such things are left a prey to [109] mice or insects. The vinegar and sauce bottles stand without the corks. Delicate things, which soon lose their fine aroma, as tea, coffee, and spices, are kept in uncovered canisters: the lid is first left off, then mislaid. The treacle jar stands open for stray fingers and flies to disport themselves therein. Capers are put away uncovered with vinegar, and when next wanted are found to be mouldy. Perhaps the juice of a lemon has been used; the peel, instead of being preserved, is thrown away, or left lying about till valueless. Herbs, which should have been at once dried and sifted, are hid away in some corner to become flavourless and dirty, and so on with every kind of store and provision.

It is impossible to calculate how many pennies are lost daily, in a large number of houses, by the absolute waste of pieces of bread left to mould or thrown out because trouble to utilise them cannot be taken. Whoever thinks anything of the small quantities of good beer left in the jug; it is so much easier to throw it away than put it in a bottle? Or who will be at the trouble of boiling up that "drop" of milk, which, nevertheless, cost a penny, and would make, or help to make, a small pudding for the next day? Then, again, how many bits of fat and suet are lost because it is too much trouble to melt down the first, and preserve the other by very simple and effectual means?

Butter in summer is allowed to remain melting in the paper in which it is sent in, or perhaps it is put on a plate, to which some pennyworths of the costly [110] stuff will stick and be lost. One would think it would be as easy at once to put it into cold salted water, if better means of cooling could not be used.

If we pause here, it is not because we have exhausted the list of things most woefully wasted, mainly from want of thought, but because we have not space to enumerate more of them. We can only add that the importance of small household savings cannot well be overrated, both because of the principle involved and because of the substantial sum they represent together. There is no need in any household for even a penny a day to be wasted; and yet if we look closely into things, how much money value is lost daily in some one or other of the ways we have mentioned. In the course of the year, the daily pennies mount up to many pounds, and we are sure that it is much safer once in a way lavishly to spend the shillings than to be habitually careless of the outgoings of the pence.

Although it is not necessary that the mistress of a household who can afford to keep servants should herself do the cooking, or spend much time in her kitchen, it is absolutely necessary that she should understand the best methods, and know how everything should be done.

Many people will say that it is unbecoming for women to be *gourmands*; we agree with them, and that it is equally unbecoming for men to be so. But to be a *gourmet* is another thing; and we ought not to lose sight of the fact that food eaten with real enjoyment and the satisfaction which accompanies a [111] well-prepared meal, is greatly enhanced in value. Professor C. Voit has clearly pointed out, in his experiments and researches into diet, the great value of palatable food as nourishment, and how indispensable is a certain variety in our meals. "We think," he says, "we are only tickling the palate, and that it is nothing to the stomach and digestive organs whether food is agreeable to the palate or not, since they will digest it, if it is digestible at all. But it is not so indifferent after all, for the nerves of the tongue are connected with other nerves and with nerve-centres, so that the pleasure of the palate, or some pleasure, at any rate, even if it is only imagination, which can only originate in the central organ—the brain—often has an active effect on other organs. This is a matter of daily experience. Without the secretion of gastric juice the assimilation of nourishment would be impossible. If, therefore, some provocatives induce and increase certain sensations and useful processes, they are of essential value to health, and it is no bad economy to spend something on them."

It is surely somewhat singular that Englishwomen, who have excelled in almost every other craft, should be remarkable for their want of skill in cookery. They have not been dismayed by any difficulties in literature, art, or science, and yet how few are there among us who can make a dish of porridge like a Scotchwoman, or an omelette like a Frenchwoman! The fact would seem to be, that educated women having disdained to occupy themselves either theo [112] retically or practically with cookery, those whose legitimate business it has been have become indifferent also. The whole aim of the modern British cook seems to be to save herself trouble, and she will give as much time and thought to finding out ways of doing things in a slovenly manner as would go to doing them properly.

No doubt cooks have often so much work of other kinds to do that they cannot give the necessary time to cooking. In a case of this kind, the mistress should herself give such help as she can, and bring up her daughters to help in the kitchen. People in middle-class life often expect the cook to do all the kitchen work, and frequently some of the house work. Of course, in small families, this is quite possible to be done, and it is always best for servants, as for other people, to be fully employed. But in large families it is impossible the cooking can be properly done, when the cook is harassed by so many other occupations. Thus, because it takes less time and attention than cooking smaller dishes, huge pieces of meat are roasted or boiled daily, and the leg-of-mutton style of dietary is perpetuated—declared to be the most economical, and, in short, the best for all the world.

Probably it is because bread and butter can be bought ready made, and involve no trouble, that they are held to be the chief necessaries of life in every English household. Some children almost live, if they do not thrive, on bread and butter. Thoughtless housekeepers think they have done their duty [113] when they have seen that a sufficient supply of these articles has been sent in from the shops. When we insist that everyone should have home-baked bread, at once we shall be met with the "penny-wise" suggestion that home-baked bread costs more than baker's, because, being so nice, people eat more of it. Good bread, we need not say, is far more nourishing than that which is made from inferior materials or adulterated even with non-injurious substances for wheaten flour. Then all the other difficulties come to the fore: cook spoils the bakings, the oven is not suitable, and so on. To all these we answer: A good housekeeper, one who looks beyond the sum total of her weekly bills, who thinks no trouble too great to provide such food as will maintain the health of her family, will have home-baked bread.

There are other points in domestic management which do not receive the attention they deserve. Of these we may cite the use of labour-saving machines and of gas for cooking.

How often do we hear it said: "I always have such and such a thing done in that way, because it was my mother's way!"

This may be very nice and very natural, but it is nevertheless a sentimental reason. What should we think of a person who insisted on riding pillion, because her mother rode pillion? Yet, this really is pretty much the same thing as we see every day, when ladies are so wedded to old ways that they persist in employing the rough-and-ready implements of domestic use, the pattern whereof has been handed [114] down from the Ark, instead of modern and scientific inventions which save both time and trouble. In no other department of the national life have the people been so slow to adopt simple machinery as in that of the household.

It is alleged, in the first place, that labour-saving machines are expensive; in the next place, that servants do not understand them, and that they are always getting out of order.

As to the first objection, we would say that as these machines — we speak only, of course, of really good machines — are made, not only with the object of saving labour, but material, the original cost of them is in a short time repaid. As regards the second objection, it seems incomprehensible that servants should not use with care and thoughtfulness machines, which not only save time and trouble, but greatly help in making their work perfect.

There is no doubt that by the more general adoption of machinery household work would be much lightened, and that if there were a demand for it, enterprise would be much stimulated, and many more useful helps would be produced. As it is, manufacturers hesitate to bring out new inventions at a great expense, when there is a doubt of securing the appreciation of the public.

Only the other day we were inquiring for a little machine we had seen years ago, and were told by the maker that, "like many other useful things, it had been shelved by the public, and ultimately lost."

Let us take the case of making bread at home. [115] By the use of a little simple dough-mixing machine, supplied by Kent, 199, High Holborn, the operation is easy, quick, cleanly, and certain. We have had one of these in use for more than ten years, and during that time have never had a bad batch of bread. Not only in this machine do we make ten to eleven pounds of dough in five minutes, but the kneading is most perfectly done, and there is the great advantage of securing perfect cleanliness, the hands not being used at all in the process. Yet we do not suppose that any number of the people who have admired the bread have set up the machine. It cannot be the cost of the machine, as it is inconsiderable, which prevents its more general use, since in households where expense is not an object the primitive process is still in vogue.

Many people imagine that washing machines are only needed in large families where all the washing is got up at home. But, if ever so small or only an occasional wash is done, there is no exaggerating the comfort and advantage of a machine which washes, wrings, and mangles. So far from injuring linen, machines of the best kind wear it far less than rough hand labour, and with reasonable care it will be found that delicate fabrics are not split in the wringing by a good machine, as they so frequently are by the hand.

Then there is the case of the knife-cleaning machine. There are families who, instead of using one, employ a boy to ruin their knives by rubbing them on a board with Bath brick. They do so, they will tell you, "because [116] machines wear out the knives." The slightest acquaintance with the mechanism of a good knife-cleaning machine should suffice to show that the brushes cannot wear out the knives, whereas the action of the board and brick is the most destructive that can be imagined. The objection of undue wear being disposed of, we are told that the machines soon get out of order, and are a constant expense. Of course, with careless usage anything will come to grief, but the fact remains that Kent, the leading manufacturer of knife-cleaners, has published a certificate from a lady who has had in constant use, for thirty years, one of his machines, which during that time has required no repairs. As to knives, we know of some which have been cleaned daily for twenty-five years in a machine, and are very little the worse for wear.

Dressmakers tell us that, but for the sewing machine, an elaborate style of trimming ladies' dresses would be impossible. We know that many inexpensive delicacies, which it is not practicable to have now because of the time and trouble they require, could easily be managed by the use of little articles of domestic machinery. For instance, take potted meat. There is the excellent Combination Mincer, also Kent's, by which this is rapidly and perfectly done, and which enables cooks to use up many scraps of material in a most acceptable way, and without the labour of the pestle and mortar. This machine, however, is but little known. It costs but a sovereign, is useful for all mincing [117] purposes, and makes the best sausages in the world.

To make sausages properly, a machine must have an adjustment of the cutters by which the sinews of the meat and bits of skin are retained on them, as nothing is so unpleasant as to find these when eating the sausages. Thus it will be seen how necessary it is, in setting up machinery which should last a lifetime, to have the best inventions in the market. Not very long ago, a friend asked our opinion on the merits of the different makers of knife-cleaning machines. We explained to her the mechanism of the best of them, pointed out the superior workmanship, and that she should not grudge the money to have one which would do its work properly and be durable. Probably under the impression that "in the multitude of counsellors there is wisdom," our friend made further inquiries, and ended by buying a much-advertised machine which, she was assured, was better and cheaper than that of Kent, the original patentee. When she had the machine home, and calculated, together with the cost of carriage, her own expenses in going to London to choose it, she found that she had saved exactly eighteenpence, and then that her bargain would not clean the knives!

The prejudices which for a long time existed against cooking by gas have gradually cleared away now that improved stoves have been introduced, and the public have experience of its many advantages. There are yet some difficulties to be met in bringing gas into more general use, one of which, the high price [118] charged for it, is beyond the control of the housekeeper, and another, that of teaching servants to be economical and careful in its use. When this last can be overcome, even with the first named drawback, gas will not be found more expensive than coal. The cost of wood, of sweeping the chimney, and the extra wear and tear occasioned by the soot, smoke, and dust of a coal fire, must be calculated in addition to the fuel itself.

It will be seen, when we say that the entire cooking for a small family having late dinners, bread baked, and much water heated, is done for something under £2 a quarter, that gas as a fuel is not so great an extravagance after all. The stove used has the oven lined with a non-conducting substance, which has the advantage of keeping the heat within instead of sending it into the kitchen, as stoves made only of iron plates are apt to do. We have but space to add that the benefit to health, the cleanliness, the saving of time, labour, and temper, to say nothing of the superiority of cooking done by gas in such a stove as has been described, can only be fully appreciated by those who, like the writer, have had twenty years' experience of all these advantages.

[119]

NEW ZEALAND FROZEN MUTTON.

The high price at which meat has stood for some years has made it necessary for the working classes to restrict themselves to a scanty allowance of animal food, and this often of poor quality. The difficulty of providing joints of meat for their families has, indeed, also been felt severely by people who are comparatively well-to-do. Under these circumstances capitalists have thought it worth a considerable investment of money to discover some means of bringing the cheap and magnificent supplies of New Zealand into the English market. After many failures, success has at length crowned the enterprise, and nothing can exceed the perfection in which New Zealand mutton is now placed on the English market. It is universally admitted that the meat, both as respects its nutritive value and its flavour, is unsurpassed, while the price is very moderate. The same remarks apply to New Zealand lamb. It commences to arrive in January, and is in the height of its season when our English lamb is a luxury which can only be enjoyed by the few.

Nelson Brothers, Limited, stand foremost among the importers of this invaluable food supply. The mutton and lamb selected by them is of the highest quality, and their system of refrigeration is perfect. In summer these New Zealand meats have a great advantage over the home supply, as although in keeping they may lose colour, they remain good and sweet much longer than English-killed meat.

The Company have large refrigerating stores under Cannon Street Station capable of holding some 70,000 sheep, and have recently erected stores of *treble that capacity* at Nelson's Wharf, Commercial Road, [120] Lambeth, wherein the latest improvements both as regards construction and refrigerating machinery have been adopted, in order to facilitate the development of the frozen meat trade.

Nelson Brothers have also Branch Offices at —

- 15*a*, Richmond Street, Liverpool.
- Lease Lane, Birmingham.

- Lawns Lane, Leeds.
- The Abattoirs, Manchester.
- Baltic Chambers, Newcastle-on-Tyne.
- Tresillian Terrace, Cardiff.

If any of our readers are anxious to try the meat, and are unable to procure it, a postcard to the Head Office, 15, Dowgate Hill, London, E.C., or to any of the Branch Offices, will at once put them in the way of carrying out their desire.

As it occasionally happens that from want of some little precaution New Zealand meat does not come to table in its best condition, we offer the following hints for the treatment of it:

Frozen mutton, like that which is freshly killed, requires to be hung a certain time—this is most essential to remember, otherwise the meat eats hard and tough—and it is important to observe, both when hanging and roasting, that it is so placed that the juice shall not run out of the cut end. Hind-quarters, haunches, and legs should be hung with the knuckle end downwards; loins and saddles by the flaps, thus giving them a horizontal position. The meat in winter should be kept in the kitchen some time before cooking, and after being exposed for a few minutes to a rapid heat in order to seal up and keep the gravy in the joint, it should be cooked rather slowly, thus taking a little more time than is usually given to English meat.

[121]

INDEX.

PAGE

- Albumen, 10

- Beef Tea, 12
  - " " as a solid, 15

- Beverages, 93
  - Badminton Cup, 94
  - Champagne Cup, 94
  - Cherry Cup, 94
  - Cider Cup, 94
  - Citric Acid, 97
  - Claret Cup, 93
  - Ginger, an Extract of, for family use, 95
  - Gingerade, 95
  - Lemon, Essence of, 97
    - " Syrup, 96
  - Lemonade, 94
    - " (a new recipe), 95
  - Milk, 96
  - Port Wine, Mulled, 94

- Blanc-mange, 79

- Cakes, 85
  - Almond Paste, 92
  - Chocolate, 90
  - Cocoa-nut, 89
    - " Rock, 90
  - Macaroons, 89
  - Pound, 87
    - " Plain, 87

- Savoy Sponge, 88
  - " " Lemon, 88
- Sugar Icing, 90

- Citric Acid, 9

- Creams, 74
  - Apricot, 76
  - Champagne, 83
  - Charlotte Russe, 79
  - Cheese and Macaroni, 81
  - Cherry, 80
  - Chocolate, 82
  - Coffee, 81
  - Fig, 83
  - Fruit, 78
  - Italian, 81
  - Lemon, 75
    - " Imitation, 76
  - Mandarin, 78
  - Orange, 76
    - " Mousse, 83
  - Oranges, Chartreuse of, 82
  - Palace, 77
  - Pineapple, 77
  - Strawberry, 75
    - " Trifle, 84
  - Syllabub, Solid, 79
  - Velvet, 80
  - Whipped, 84

- Essences —
  - Almonds, 9
  - Lemon, 9
  - Vanilla, 9

- Extract of Meat, 10

- Fish, Little Dishes of, 22
    - Cod Cutlets, 26
    - Eels, Collared, 30
    - Fish, Galantine of, 28
    - Herrings, Fried, 27
        - " Rolled, 27
    - Sole, Filleted, 24
        - " Fillets of, en Aspic, 29
            - " " Fried, 25
            - " " Sautés, 25
            - " " with Lobster, 25
        - ", Fried, 23
    - Whiting, Baked, 26

- Flummery, Dutch, 72

- Gelatine, 13
    - " How to use, 64

- Housekeeping, Hints on, 105

- Jellies, Nelson's Bottled —
    - Calf's Foot, 8
    - Cherry, 8
    - Lemon, 8
    - Orange, 8
    - Port, 8
    - Sherry, 8

- Jellies, Tablet, 8

- Jelly-Making, On, 61
    - Jelly, Apple, 69
        - " Aspic, 72
        - " Brilliant, 65
        - " Claret, 67
        - " Cocoa, 68

- " Coffee, 68
- " Orange Fruit, 69
- " Oranges filled with, 69
- " Ribbon, 66
- " Strengthening, 71
- " with Fruit, 66
  - Jelly-bag, how to make a, 73

- Jelly-Jubes, 10

- Lemon Sponge, 9, 70

- Lozenges—
  - Gelatine, 9
  - Licorice, 10

- Macaroni, etc., 98
  - Canapés au Parmesan, 101
  - Cheese, Boiled, 101
  - Eggs, Scalloped, 102
  - Macaroni Cheese, 99
    - " Stewed, 100
    - " Sweet, 100
    - " with Bacon, 99
      - " " Cheese, 98
      - " " Onions, 100
      - " " Tomatoes, 100
  - Mushrooms with Cream Sauce, 103
  - Rice, to Boil (a black man's recipe), 103
  - Rice with Parmesan Cheese, 101
  - Scotch Woodcock, 102
  - Vegetables, to Mince, 104

- Meat, Little Dishes of, 31
  - Brain Fritters, 35
  - Chicken, Brown Fricassée of, 42
  - Chicken Sauté, 43

- " in Aspic Jelly, 36
- Croquettes, 44
- Curry, Dry, 44
- Kidneys, Broiled, 39
  - " Sautés, 37
  - " with Mushrooms, 38
  - " with Piccalilli, 39
- Lamb's Fry, 40
  - " Sweetbreads, 41
- Marrow Toast, 35
- Meat Cakes à l'Italienne, 45
- Mutton, Cold, Potted, 33
  - " Collops, 33
  - " Cutlets, 31
  - " Pies, 34
  - " Roulades of, 32
  - " Sauté, 33
- Ox Brain, 34
- Pork Pie, Raised, 46
- Potato Hash, 43
- Sausages, Pork, 47
- Veal à la Casserole, 41
- Veal and Ham Pie, 47
- Veal Cutlets in White Sauce, 37

- Puddings, 50
  - Apple Fool, 59
    - " Meringue, 60
  - Baden-Baden, 80
  - Brandy Sauce, 53
  - Cabinet, 53
  - Capital, The, 57
  - Cheesecake, Welsh, 58
  - Chocolate, 56
  - Cocoa-nut, 56
  - Compote of Apples with Fried Bread, 59
  - Compote of Prunes, 60
  - Custard, 50

- Duchess of Fife's, 58
- Fritters, Italian, 58
- Jubilee, 55
- Natal, 55
- Omelet, Friar's, 58
  - " Soufflé, 52
- Pears, Stewed, with Rice, 60
- Queen's, 56
- Raspberry and Currant, 57
- Soufflé, 51
- Sponge Soufflé, 53
- Vanilla Rusk, 54
- Warwickshire, 54

- Soups, 11, 14
  - Artichoke, Brown, 19
  - Beef and Onion, 14
  - Beef, Lentil, and Vegetable, 15
  - Beef, Pea, and Vegetable, 15
  - Glaze, 21
  - Gravy, 21
  - Hare, 17
  - Julienne, 16
  - Mulligatawny, 18
    - " Nelson's, 14
    - " Thin, 18
  - Rabbit, Brown; Clear, 17
  - Turtle, 19
    - " Mock, 21
  - Vermicelli, Clear, 16

- Tinned Meats, 12

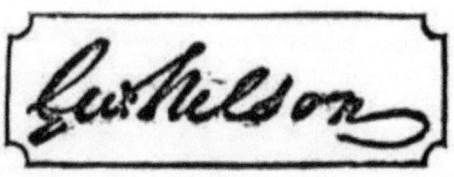

CHARLES DICKENS AND EVANS, CRYSTAL PALACE PRESS.

By Royal  Letters Patent.

For First Class Jellies

NELSON'S
OPAQUE GELATINE
SHOULD ALWAYS BE USED.

See Recipe, Page 65.

NELSON'S
TABLET JELLIES.
*Orange, Lemon, Calf's Foot, Cherry, Raspberry,
Vanilla, Apricot, Pear, Apple, Black Currant,
Pine Apple, Noyeau, etc.*

Quarts, 9d.; Pints, 6d.; Half-Pints, 3d.

WINE TABLET JELLIES.

Port, Sherry, Orange.

Pints only, 9d.

These new Jellies are perfectly pure and wholesome, and the flavours excellent, while their exceeding cheapness brings them within the reach of all classes.

G. NELSON, DALE, & CO., Ltd., 14, Dowgate Hill, London.

By Royal  Letters Patent.

NELSON'S SOUPS.

These Soups are already thoroughly cooked and seasoned, and can be prepared for the table in a few minutes.

BEEF AND CARROTS.
BEEF AND CELERY.
BEEF AND ONIONS.
MULLIGATAWNY.

In Pint Packets, 6d. each.

BEEF, PEAS, AND VEGETABLES.
BEEF, LENTILS, AND VEGETABLES.

In Quart Packets, 6d. each.

Penny Packets of Soup for charitable purposes.

NELSON'S
EXTRACT OF MEAT,
FOR MAKING AND IMPROVING
SOUPS, GRAVIES, BEEF-TEA, etc., etc.

In Ounce Packets, 4d. each, and 1 lb. Tins, 5s. each.

NOTE.

One packet is sufficient for a Pint of Strong Soup.

G. NELSON, DALE, & CO., Ltd., 14, Dowgate Hill, London.

By Royal  Letters Patent.

NOTICE.

On RECEIPT of POSTAL ORDER for 2/6
A BOX CONTAINING SAMPLES OF

NELSON'S SPECIALITIES
AND A COPY OF
"NELSON'S HOME COMFORTS,"
Will be sent, CARRIAGE PAID, to any address
in the United Kingdom, by
G. NELSON, DALE, & CO., LIMITED,
14, Dowgate Hill, London, E.C.

May also be obtained through any Grocer at the same price.

N.B. — A Copy of "Home Comforts" will be sent, gratis, on receipt of Penny Postage Stamp.

G. NELSON, DALE, & CO, Ltd., 14, Dowgate Hill, London.

www.ingramcontent.com/pod-product-compliance
Lightning Source LLC
Chambersburg PA
CBHW031436210526
45464CB00005B/2225